THE UNITY OF THE SCIENCES
VOLUME ONE

THE QUANTUM FOUNDATION OF THE BIOLOGICAL SCIENCES

To my
True Parents

With Thanks.
.

The Unity of the Sciences
    Volume One: Quantum Foundations of Biology
    Volume Two: Mathematics, Physics and Chemistry
    Volume Three: Life, Mind and Spirit

ISBN: 978-1-304-53708-9

The Quantum Foundations of the Biological Sciences
    Firtst Edition 2013

Originally published as "Do Prpteins Teleport in an RNA World?
    Second edition 2012
    First edition   2005

All right reserved
© 2013 Richard Lewis

RICHARDLLL@MAC.COM

THE UNITY OF THE SCIENCES
VOLUME ONE

THE QUANTUM FOUNDATION OF THE BIOLOGICAL SCIENCES

By

Richard L. Lewis, PhD

UNIFICATION THOUHT INSTITUTE
SEOUL • TOKYO • NEW YORK

Contents

Unity of the Sciences? 1

The Reluctant Revolution 11

Quantum Probability 30

Quantum Interaction 42

Quantum Probability Forms 68

Operating System of Life 100

Eden, Wombs and Beds 110

Quantum Adam & Eve 114

Appendix 146

UNITY OF THE SCIENCES?

Science is, almost indisputably, one of the most positive of the remarkable developments that have emerged in the last 500-or-so years since the Renaissance. The fruits, for good or ill, of scientific insights into how the world actually works have earned its practitioners a magisterial authority reserved in earlier ages for the revealers of mystical truth.

A bedrock belief of all the sciences—it can be considered the basic philosophical prerequisite for a discipline to be counted as a science—is that there is an objective reality "out there" to be studied. Moreover, it is the same objective reality for all of us. The holy grail of science is to come up with an accurate description of this objective reality.

While words can do a lot, the most accurate descriptions in science are couched in terms of mathematical shorthand. For example, two key insights by Newton and Einstein are succinctly described as:

$$E = mc^2 \quad F = ma$$

Unless we have to, however, we will try to stick to words to get the point across.

HIERARCHY OF SCIENCE

To those unacquainted with its inner workings, scientists can seem to be a part of a vast, monolithic entity—an almost-priesthood with magic powers (and possibly-suspect motives, as attested by the plethora of evil-scientists with British accents in the movies).

To the many workers focused on the endless developments within their own subspecialty of a science, however, science seems less a unified entity than a multitude of relatively independent disciplines:

"The statement 'chemistry and biology are branches of physics' is not true. It is true that in chemistry and biology one does not encounter any new physical principles. Nevertheless, the systems on which the old principles act differ in such a drastic and qualitative way in the different fields that it is simply not useful to regard one as a branch of another. Indeed the systems are so different that 'principles' of new kinds must be developed...."[1]

For all this sense of independence, however, the autonomy of each discipline to develop its own conceptual framework is constrained by the pecking order in science. The rule is simple: a scientist is free to construct any theory so long as it does not contradict what has been established as an accurate description at a lower level in the hierarchy. The chemist is not free to contradict the concepts of physics, the biochemist must respect the rules of the chemist, and a biological theory cannot contradict biochemistry. For example, while neurologists have great latitude to develop concepts to explain the phenomena they encounter in the brain, they are not free to contradict the principles of cell interaction established in biology. Similarly, an evolution theorist cannot contradict the principles of biology—evolution depends on biological processes.

A scientist who wishes to excel at a discipline needs, at the minimum, to have a good grounding in the discipline just below: the evolutionist must know his biology; the chemist his physics. This is a one-way street, however, for you do not need to know anything about the levels above to do well in a discipline. A physicist can excel without knowing any biology

---

[1] H. M. Georgi, "Grand Unified Theories," in The New Physics, ed. Paul Davies, Cambridge University Press, NY (1989), p. 448.

whatsoever, for instance, which might explain the dearth of quantum concepts in mainstream genetics and the development of the body let alone evolution and the workings of the nervous system.

The physicists have no one beneath them in the hierarchy to acknowledge; their only constraint is that their theoretical constructs should be mathematically sound or, better yet, "elegant." To paraphrase a well-known eminence's stinging rejection of an aspirant's theory: It is so mathematically ugly that it is not even wrong!

Just why mathematics—a construct of human minds over many centuries—should have this uncanny ability to describe the natural world so accurately is not at all clear. "Opinions range from those who maintain that human beings have simply invented mathematics to fit the facts of experiment, to those who are convinced that there is a deep and meaningful significance behind nature's mathematical face."[1]

Mathematics, of course, is much broader than just its descriptive role in science and can describe constructs that have no—so far—use in describing objective reality, the constructs in nature.

Mathematics is also self-contained; it has nothing more basic beneath it (except a faith in logic).

Whatever the rationale; all scientists aspire to put their disciplines on a firm mathematical foundation—to be a "hard" science—rather than being vague and suggestive—to be second-classed as a "soft" science. To have to resort to vague and shifting English etc. words and, a sure sign of fluffiness, endless hand-waving.

A simple analogy to the hierarchical nature of science is the Empire State Building just blocks from where I have worked for a score of years. The foundation, the basement is fundamental physics. Up go the floors, blending into chemistry then biochemistry then genetics then development to the floors in the 100s dealing with evolution, brain function etc.

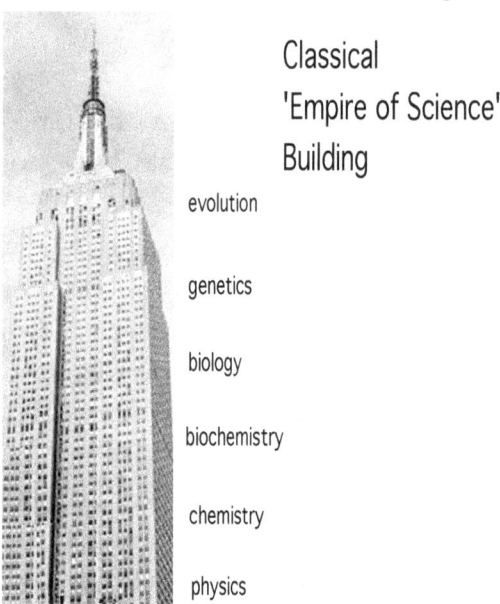

Classical 'Empire of Science' Building

evolution

genetics

biology

biochemistry

chemistry

physics

TWO SCIENCE FOUNDATIONS

Newton is rightly considered the Father of Science as we know it. The themes he developed in classical physics have appeared throughout the scientific structure. Therefore, while biology might not be a branch of physics, the basic Newtonian concepts of classical science permeate biology.

Of course, one is philosophically free to drop the hierarchical constraints in constructing a theory of how the world works; but the construct will be something other than science as it is practiced today. The classic historical example of this is

---

1 Paul Davies, The Mind of God: The Scientific Basis for the Rational World, Simon & Schuster, NY (1992), p. 140.

the attempt to explain living systems by the introduction of a "vital force" in one guise or another. While there are many philosophical constructs that embrace this as an acceptable explanation, none of them are part of biology because particles, atoms and molecules can be understood without a vital force and, if electrons and quarks don't have it, neither do the atoms and molecules they comprise, nor do cells or higher organisms.

On the other hand, we can expect the converse to be true. If particles, atoms and molecules have some aspect essential to their structure and function, then we might expect some biological systems to involve this aspect as well.

While all scientists accept this pecking order, there are currently two quite different physics to be found at the foundations of the scientific edifice.

The conceptual framework in which physics started out, and the one that is still used in the biological sciences, is described by many adjectives: Newtonian, classical, nineteenth-century, old-fashioned, high school, etc.

Classical physics, however, has been completely replaced by the quantum revolution. For the classical worldview was found to be almost totally inadequate. The more sophisticated replacement framework, the one physics currently embraces, is also multi-monikered: post-Newtonian, New Physics, quantum mechanics, twentieth-century, modern, post-grad, etc.

The new physics is based on the theories and explanations of quantum physics that successful explained a wide variety of phenomena that the old physics was incapable of dealing with. We will list these shortly.

The quantum perspective now pervades all of physics and it has been remarkably successful in dealing with things as different as the first microsecond of the Big Bang and the workings of lasers and superconductors.

The remarkable success of the new physics makes it unlikely that its concepts will be completely replaced by future theoretical developments. It is, of course, possible that they will suffer the same fate as the Newtonian concepts—and they were equally successful in their own day—and turn out that they are phenomena of a much deeper and sophisticated reality.

The new physics is, indeed, so radically weird to the classical mind that it is very difficult to accept the basic concepts at face value As one wit put it: not only is reality stranger than you think, it is stranger than you can think. And we are stuck with the weird quantum view which has gone from one success to the next throwing off a plethora of goodies based on electronics such as my Mac with its laser-run CD burner and DVD reader.

"Perhaps, someday, an experiment will be performed that contradicts quantum mechanics, launching physics into a new era, but it is highly unlikely that such an event would restore our classical version of reality. Remember that nobody, not even Einstein, could come up with a version of reality less strange than quantum mechanics, yet one, which still explained all the existing data. If quantum mechanics is ever superseded, then it seems likely we would discover the world to be even stranger."[1]

Therefore, science, at the commencement of the third millennium, is not just multi-disciplinary; it is a discipline with something of a split personality. In the hierarchy of physics, chemistry, biochemistry, biology and evolution, the switch-over from one science system to the other is to be found somewhere between physical and biological chemistry.

So, while the biology of our era is proud of its firm foundations in the "hard" sciences (those amenable to mathematical rigor), the physics in which it is rooted is the classical physics of Darwin's day. "It is most ironic that today's perceived conjunction between physics and biology, so fervidly embraced by biology in the name of unification, so deeply entrenched in a philosophy of naive reductionism, should have come long past the time when the physical hypotheses on which it rests have been abandoned by the physicists."[2]

---

1 Johnjoe McFadden, Quantum Evolution, W. W. Norton, NY (2000), p.219

2 Robert Rosen, Life Itself, Columbia University Press, NY (1991), p. 18.

  Quantum 'Empire of Science' Building

There is still, of course, the sense that science should be a unified structure: "How does nature encompass and mold a billion galaxies, a billion, billion stars—and also the earth, teeming with exuberant life? New insights into how nature operates come from parallel advances in particle physics and in molecular biology; advances that make it possible to examine fundamental physical and biological processes side by side. The resulting stereoscopic view deep into the past reveals a previously hidden, unifying logic in nature: its paradigm for construction."[1]

This is the task of this work, to establish the basic quantum principles, based on the new physics, which are applicable to all levels of the scientific edifice.

On a personal note: I have been fascinated by all the sciences since early schooldays and chose the interdisciplinary biochemistry for my graduate education. (Like my inspiration, Isaac Asimov.) When it came to choose the topic for my Ph.D. thesis I came up with "The Impact of the Quantum Revolution on Evo9lutionary Thought." To both my and my advisor's surprise, I could not find any impact. The change in the basement had yet to be communicated to the top floors.

I wrote this somewhat 'negative' thesis and my late advisor encouraged me to expand it into a book. The book you are reading more than twenty years on.

To say that pre-twentieth century scientists were content with Newtonian physics, chemistry, etc. is an understatement. One eminence, commenting on the state of classical science at the end of the nineteenth century—at its apogee just before the 'unexplainable weird' became apparent—declared that all that now remained was mopping up, getting ever-increasing accuracy and more and more decimal places. He was oh-so wrong.

Scientists were, almost literally, dragged kicking-and-screaming into accepting the quantum worldview because the only deity in science insisted upon it. That deity is experiment. For no matter how elegant, mathematically-sound, politically-correct, etc. a theory might be, if it contradicts experiment it is crumpled up and thrown into the wastebasket.

BASEMENT PROBLEMS

Genetics has been called the Plastics![2] of our age. For the science of genetics is still in a state as alchemy was to current chemical prowess. The DNA-protein connection was established just a half-century ago, and the possibilities that are opening up, even with our primitive understanding, seem endless.

Just a few possibilities are:

In the near future, repairing genetic defects, therapeutic cloning, ordering up a 20-year younger twin, etc. In decades: designing one's children, artificial wombs—and if we are not wise, all the monstrosities that strife can give birth to.

---

[1] Edward Rubenstein, "Stages of evolution and their messengers," Scientific American (June 1989), p. 132.

[2] Whispered to Dustin in The Graduate, 1960s.

Who can tell where we will go with genetic engineering as the technologists move in behind the conceptual advances in understanding. One thing is certain, however; there are many Nobel Prizes and mega-dollar IPOs waiting for plucking. And lawsuits; and laws being fiddled with.

Somewhat spoiling this triumphal, exponential advance, however, is a grubby little secret: The conceptual edifice being constructed by the geneticists is lacking a solid foundation. This is nothing to do with the glamorous DNA, which gets all the press, but down in the very the basement of genetics, the realm of the proteins. Proteins lack the glamour of DNA, yet they do almost all the actual work.

If nucleic acids are the white-collar hierarchy on the upper floors, then proteins are the blue-collar handymen from the basement.

The management of even the most complex of organisms is founded on this sequence of cause-and-effect in the bottommost basement of our Empire State of the life sciences:

> *Higher control levels release patterns from DNA onto RNA which is translated into a linear chain of aminoacids which folds and compacts to a protein with an active site that fits a molecule.*

PROTEIN FOLDING

All but one of these steps are well understood, leaving just the "protein folding" step as a major mystery 50 years into the genetic revolution. Protein folding is the technical term for the last step in making an active enzyme, for example. For all proteins are first spun out as a long, sticky thread that has to fold up into the precise 3-D shape. The precise shape that is the active protein.

In more complicated situations, it seems that the chains have to interact with other proteins to fold correctly.

Even the simple, unaided situation is, however, a puzzle. Scientists have already taken into account all the known interactions such as hydrogen bonding, hydrophobic & hydrophilic interactions, 'metal–ion chelation', and 'steric hindrance' and calculated the predicted forms. But here, so far, they have hit a snag. The problem is that: "calculations designed to predict the three-dimensional structure of proteins ... invariably give far too many solutions. In the literature on protein folding, this is known as the 'multiple-minimum' problem."[1] There are so many solutions it would not be possible for a protein to test all of these until it finds the right one, it would take too long. A small chain of 150 amino-acids testing 10^{12} different configurations each second would take about 10^{26} years—a billion, billion times the age of the universe—to find the 'correct' configuration. Yet the refolding of a denatured enzyme takes place in less than a minute.

Naturally, this problem has attracted the attention of many workers. A recent review of advance in this field noted, "It is not yet possible to predict a three-dimensional structure from just the amino-acid sequence, except by homology with a protein of known structure. Nevertheless, understanding the basic rules of protein architecture is now well advanced, and it is becoming possible to design folded structures de novo."[2]

While our understanding of the internal systems involved in protein folding is currently minimal. one thing is very clear: they all involve linear chains of amino acids. Occasionally a chain will be linked in a circle, even more rarely a peptide side

---

[1] R. Sheldrake A New Science of Life , Tarcher (1981) p 70

[2] T. E. Creighton, "The Protein Folding Problem," Science 240, 1988, p. 267, 240.

chain will hang off the main chain. But in large, all the peptides and proteins of life are linear chains. Admittedly these chains are often linked, but the bonds linking them are not peptide bonds (well, perhaps rarely) but the thioesters bond involving the rather unusual sulfur - sulfur bond. The problem with such linearity is that amino acids are just as likely to form branches with their side chains—quite a few have amino or carboxyl groups on their side chains and these can participate in peptide bond formation.

This is exactly what happens in natural metabolism. In an environment that favors the peptide bond, a mix of amino acids will form all sorts of branching chains as the side groups participate in the peptide bond forming. Such tangles of non-linear chains are called proteinoid.

Even random linkage of amino-acids can produce molecules with interesting properties (such as the microspheres of Sidney Fox and his collaborators) such as "catalytic activity, membrane-like properties, electrical activity, sensitivity to light…"[1]

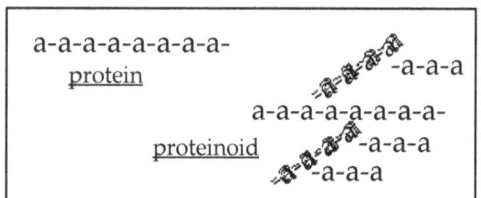

Proteinoid is a not-unlikely product of natural metabolism and some workers have proposed it as being central to proto-metabolism. If so, however, then the proteinoid has left about as much fossil evidence as has clay, namely very little.

A familiar example would be folding a plane sheet of paper into an intricate origami bird. There are a lot of steps that have to be done right to make it happen. In the same way, a nascent protein chain has to fold naturally (and usually without assistance) and properly in the same way to get to the desired end, the active form.

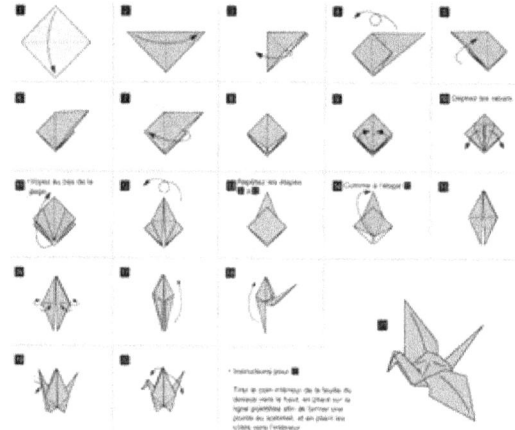

Unfortunately for classical science, there is no well-accepted explanation of just how a linear chain of aminoacids folds precisely and quickly into its active form.

Many protein enzymes can be reversibly unfolded, or denatured, by elevated temperatures (a boiled egg is irreversible denaturation). Warm the enzyme solution and the aminoacid chain unfolds—it returns to the unfolded form of its ribosomal nativity. The enzymatic activity totally disappears. Cool the solution, and the enzyme is reborn; the chain refolds into the exact same form as it had before and the enzymatic activity fully returns.

Now, while this does not sound too mysterious, in the conceptual framework of classical science it verges on the miraculous.

---

1 Shapiro, R. (1985) "Origins: A Skeptic's Guide to the Creation of Life on Earth" Simon & Schuster, Inc. NY p. 195

AMINOACID DESIRE

In order to give a broad overview of protein folding, I am going to resort to anthropomorphism—it makes things so much simpler to explain without having to use technical jargon. (If this gets irritating, just mentally translate "desire" into high energy, low probability state; and "mutual satisfaction" into bound, low energy, high probability state.)

Each of the 20 varieties of aminoacids has a set of 'desires' it seeks to 'satisfy' chemically. In the natal, extended state, each one of the aminoacids in the chain clamors and insists on satisfying its needs with a complementary partner or a ménage-a-many: Positive seeks negative charge; Water-hater seeks same for deep dehydration; Water-lover seeks ice princess; Active hydrogen-bonder desires passive partner; Sulfur looking for same to cohabit; Any aromatics out there? Etc. etc.

Some aminoacids have many needs; some have just one; some are complex and massive, others simple and small. Some are strident in their demands while other are moderate in their requirements. Odd proline has a kink, two cysteines like to cross-link, while glycine, the simplest, makes no demands at all.

There are a set of different properties that the amino acids have in different amounts:

Some participate and encourage the natural tendency of the backbone to wind up in an alpha helix, while others hate to do this and prefer straightness in their neighborhood. There is also another way the chain can fold, the beta sheet in which chains lie parallel to each other, some One love this, some do not. Proline has a kink in it making the turns in beta sheets and ending any alpha-helix. The 'blanks' are easy either way; they can be straight or happily dance a helix or pleat a sheet.

Water: Some aminoacids are very good at providing forms for water to participate in while others provide only 'hostile' forms that repel water. Some are acid, some are basic. Some are strong acids and or bases whose charges have to be satisfied. Some are H-bond donors, others are acceptors. Cysteine has a -SH group at the end of a hydrocarbon string. This likes to bond with a -SH on another chain forming a disulphide bond, –SS– and so "cross links" chains. (Insulin, a familiar protein, has four chains all linked by such 'disulphide' linkages. Aromatic aminoacids have bulky rings that are most comfortable when they can stack together like pancakes.

Each aminoacid can find satisfaction with many partners, i.e., they are promiscuous, or perhaps better put, generalists. Some swing both ways, especially where water is concerned. Aminoacids will accept anyone with the right charms.

A similar list of 'desires' for the nucleotides is much simpler. They are the opposite of the generalist aminoacids. Only one partner, and one only, will satisfy a nucleotide's monogamous desire.

In the following discussion, DNA is going to lose some of its star power. In fact, we will hardly mention it at all. Rather, our focus will be on RNA in all its many guises.

Only two chemical differences distinguish RNA and DNA, and both serve to make DNA more inert and long-term stable than RNA (suitable for shipping down the generations). The two differences are

The ribose backbone in DNA lacks a hydroxyl and its h-bonding ability. One of the four nucleotides has one extra oily-spot, or CH_3– radical added to it.

DNA has less tendency to interact with water and a greater tendency to self-interact than DNA and is a lot more stable and inert. As we shall see, DNA plays a role similar to the shiny CD that I received from Microsoft with Office 2004 on it, and the Word that I am creating this book with.

In and of itself, it is rather boring. It is inert, which is an excellent trait for something being sent from Redmond down the somewhat hostile environment that is the US Postal Service.

Insert it into the Mac, however, and it springs to 'life.' The stored program becomes an active program, a sophisticated linear construct running on the operating system and blossoming into the faultless writing tool that is this Word. (Yes, I am looking for financial backing from a generous sponsor and for such am willing to overlook my long, bumpy and expensive history with MS Word since v1.0, and the tortuous separation from my beloved, elegant, slim 5.1a, now just a memory.)

While DNA is as the compact disk, RNA is the program, the operating system, and the PowerPC chip. The rest of the computer is protein.

Therefore, the focus will be on RNA. In fact, while all of the RNAs we will encounter will be copied off a DNA, I will probably forget to mention it and take that fact for granted.

Where DNA has a T, RNA has a U. I will generically use U, even when discussing DNA, to simplify things.

Each of the four nucleotides, N, has a complement N̲ that it will avidly embrace, while it is actively repelled by the other three that are not its one-and-only.

Except for protein-mimics that can pry them temporarily apart as in duplication and transcription—no other partner will do for a nucleotide. Unlike the aminoacids, the nucleotides are picky specialists. Actually, proteins acting as nucleotide mimics, do pry apart their relationships apart temporarily such.

Back to proteins. Constraining the possible hook-ups is the chain that binds them. The desires of the needy-neighbors intrude and have to be taken into account—compromises have to be made. Moreover, the chain itself has needs: it's happy to self-interact into coils and sheets if given a little encouragement.

There are also a multitude of water molecules enveloping and interacting with the chain; and water molecules have a driving need to be as ice-like as possible. While individually small, their overwhelming numbers make them major players in the final configuration.

The final, unique configuration that the chain folds into maximizes the overall satisfaction of almost everyone: the aminoacids, the chain as well as the clinging coat of water. Each of the billions of identical chains folds and compacts to exactly the same configuration, the 'active' form of the protein.

As in politics, however, not every constituent can be satisfied—the best possible compromise can still leave a few aminoacids frustrated. These unsatisfied few, the excluded-from-the-party aminoacids, end up having to seek chemical satisfaction with the multifarious molecules in the milieu about them.

This frustrated minority is the source of the catalytic, manipulative abilities that characterize proteins. In the classical worldview, we have the 'lock-and-key' metaphor to guide us: we can think of the aminoacids as having "bumps and hollows" that fit together like, yes, lock and key.

In a few proteins, such as the albumin in egg white and blood, almost every aminoacid is happy, and the protein is inert. This is as close as living systems get to storing aminoacids.

Calcium flip

Less spectacular, but of tremendous importance for the later discussion, is that a folded protein is not a static thing. It is dynamic and can change abruptly.

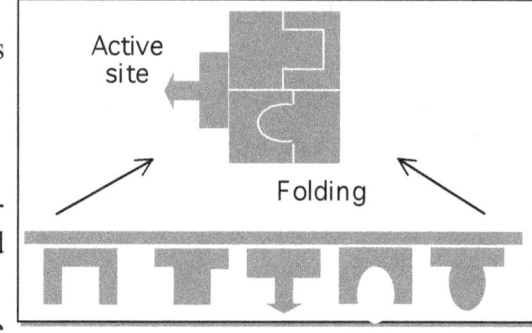

A common example involves the calcium ion, a tiny but intense source of positive charge. Normally the concentration of calcium is kept extremely low inside a cell, while it is high outside the cell wall. This means that every single aminoacid chain, fresh off the ribosome, has folded in the total absence of calcium ions.

Almost all cells are sensitive to being prodded in a way that is of particular importance to the cell. When the sentinels in the cell wall receive this important message, they open the gates and allow calcium to flood into the cell interior.

For most proteins, this is of no consequence. But for dozens of proteins it matters a great deal. When calcium appears, their current form suddenly becomes improbable and a configuration including calcium becomes very probable. The chain jumps to this new form including calcium. (In our anthropomorphism: a gorgeous woman arrives and relationships shift and a new balance is reached.)

Unlike the original configuration, this calcium-plus form has a very active site. It immediately gets to work as a pebble starts an avalanche, and the whole cell is quickly informed that the summons has come. Dozens of different processes are hit on by the rapidly-activated horde and the cell 'responds' to the important signal received by the cell wall sentinel.

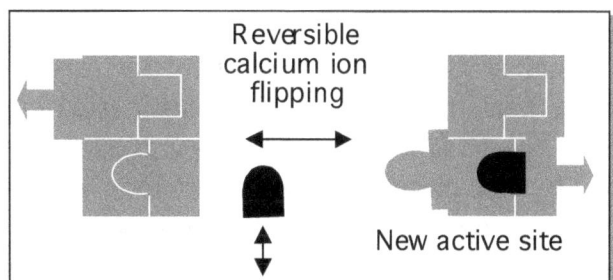

This is basically, what the muscle cells are doing in my typing fingers. A muscle cell is jolted awake by a neuron; calcium floods in, the muscle proteins flip to the short form; the cell contracts; my finger moves, the calcium is rapidly pumped out; the proteins flip back to the long form; the cell relaxes and awaits the next jolt from my brain.

Dozens of times a second, back and forth the aminoacid chain flips from one distinct form to the other. Clearly, whatever the process is by which the chain finds its final form is very fast acting indeed.

The same type of argument can be applied to the 'folding' of nucleic acid strands except that the 'needs' of nucleotide bases are singular and fussy: they only find satisfaction with their complementary base: no other. Aminoacids are generalists; nucleotide bases are specialists.

In addition, while protein folding (usually) involves the chain collapsing in upon itself, the "folding" of nucleic acids (usually) involves folding by lining up with another chain. Just as quickly as does a cooling protein fold, so do multi-thousand strands of nucleotides align with their complements in a cooling solution and coil up neatly in a double helix.

The DNA helix is actually quite dynamic and can be profoundly altered by inviting such things as proteins or testosterone into its configuration.

The Classical Commute

In classical science there is a concept that is taken for granted; it is so commonsensical that you undoubtedly agree with it. This 'belief' is that in order to go from point A to point B you have to cover all the points in-between.

This classical concept implies that the chain smoothly and continuously writhes and twists around before it settles into the 'correct' configuration. A newly-minted chain moves and twists about, testing the possible configurations for overall satisfaction, before settling down into the configuration that makes everyone happy.

In the conceptual framework of classical science (the one taught as "science" in high school) there is no way around it: each aminoacid is going to have to physically move—dragging the chain along with it—to each of the other aminoacids in turn to check out the possibilities of a liaison. And the aminoacids do not politely take turns—they are all actively hunting for satisfaction at the same time tugging at the neighbors to follow.

Now there are hundreds of aminoacids in a typical protein. Clearly there are a lot of different configurations that are possible, each with its associated level of overall satisfaction. So how does the chain find the best route from unfolded to fully-folded?

Taking all these aspects into account, classical physics allows us to estimate how long a simple enzyme should take to fold from the extended configuration into its unique, active form.

The result of this calculation is an eon upon eon of years measured in numbers with hundreds of digits (a million has just six, the age of the universe has just ten.)

The problem is that: "calculations designed to predict the three-dimensional structure of proteins ... invariably give far too many solutions. In the literature on protein folding, this is known as the 'multiple-minimum' problem."[1]

If the aminoacid chain has to find the quick route through a vast "configuration space" it should, on average, take almost forever to do it. Yet, the actual time taken by proteins to correctly fold into their active, compact form is measured in fractions of seconds. In the reversible denaturation we discussed earlier, all the trillions upon trillions of identical chains, on cooling, fold into the active form very quickly. Yet, theory predicts a google of years for just one to make it. And, as my fingers are typing, the muscle proteins are happily flipping from short to long to short again in milliseconds.

Quite a failure of theory!

Time for another metaphor: that of the jigsaw puzzle. Take an assembled 100-piece, chunky, wooden puzzle and attach all the pieces to a length of string. Now break it up and agitate vigorously. Time how long it takes for the puzzle to reassemble. Common sense tells us not to wait up; the chance of spontaneous reassembly, while possible, is so utterly improbable as to make winning MegaMillions look like a sure thing. The technical name for the mathematical treatment of these combinatorial possibilities is called the Traveling Salesman Problem, which sounds like a joke but is considered a serious field of study.

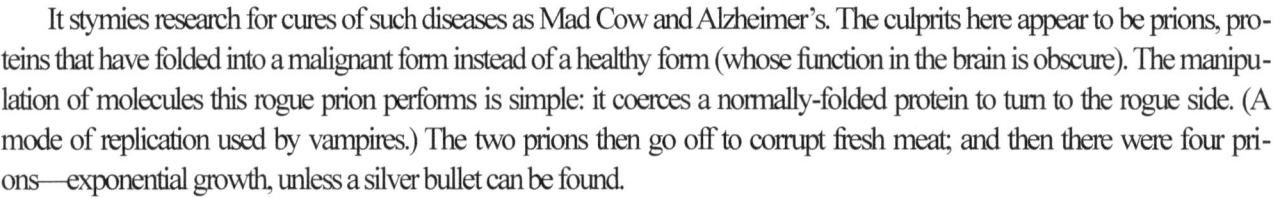

Yet this, in essence, is the best suggestion that classical science can come up with to explain protein folding.

There are problems with this scientific enigma:

It is demeaning-to-the-trade for the scientific community not to understand such a key step at the very foundations of genetics.

It stymies research for cures of such diseases as Mad Cow and Alzheimer's. The culprits here appear to be prions, proteins that have folded into a malignant form instead of a healthy form (whose function in the brain is obscure). The manipulation of molecules this rogue prion performs is simple: it coerces a normally-folded protein to turn to the rogue side. (A mode of replication used by vampires.) The two prions then go off to corrupt fresh meat; and then there were four prions—exponential growth, unless a silver bullet can be found.

A possible solution is provided by the revolution that occurred in physics starting 100 years ago.

In the next few sections, we will take a look at quantum physics and some of its implications for the rest of the sciences.

Finally, we will return to protein folding, quantum concepts in hand, and suggest a mechanism that can be experimentally tested.

---

1 Rupert Sheldrake, A New Science of Life: The Hypothesis of Formative Causation, J. P. Tarcher Inc., Los Angeles, distributed by Houghton Mifflin Co., Boston (1981), p. 70.

THE RELUCTANT REVOLUTION

The conceptual framework with which physics started out, and the one that is still in use in the biological sciences, is described by many adjectives: Newtonian, classical, nineteenth-century, old-fashioned, orthodox, conventional, etc. It can be epitomized for T-shirts as: "All is matter in motion responding to forces."

This conceptual framework was abandoned—except as a useful, if gross approximation—with great reluctance in last century because it was found to be <u>utterly,</u> and <u>totally,</u> inadequate. The more sophisticated replacement, the one physics currently embraces, is also multi-monikered: post-Newtonian, New Physics, quantum mechanics, twentieth-century, modern, way-out, totally weird, etc.

The search for a more comprehensive explanation that could deal with the experimental challenges to the classical view took physicists deeper into the nature of objective reality.

"In a sense, the difference between classical and quantum mechanics can be seen to be due to the fact that classical mechanics took too superficial a view of the world: it dealt with appearances. However, quantum mechanics accepts that appearances are the manifestation of a deeper structure ... and that all calculations must be carried out on this substructure."[1]

The new physics reached its apotheosis in the "adding endless little arrows" over-history' methodology perfected by Richard Feynman. This perspective is also called quantum electro-dynamics (QED), the official name for the theory that describes the behavior of electrons and photons in terms of internal probability.

QED is extraordinarily successful and accurate. Feynman has modestly stated that: "The theory of quantum electrodynamics has now lasted more than fifty years and has been tested more and more accurately over a wider and wider range of conditions. At the present time, I can proudly say that there is no significant difference between experiment and theory! ... To give you a feeling for the accuracy [of the quantum description of the electron]: if you were to measure the distance from Los Angeles to New York to this accuracy, it would be exact to the thickness of a human hair. That's how delicately quantum electrodynamics has, in the last fifty years, been checked—both theoretically and experimentally."[2]

The concepts and theories of quantum physics are so exquisitely successful in dealing with such a wide range of phenomena—including the furnace of the Big Bang, the graceful aging of our sun, the nature of the elements, and the workings of DVDs—that they have no serious contender.

The success of the new physics makes it unlikely that its concepts will be completely replaced by future theoretical developments. It is, of course, possible that they will suffer the same fate as the Newtonian concepts—and they were equally successful in their own day—and turn out that they are artifacts of a much deeper and sophisticated reality.

Quantum physics also graciously explains why treating atoms as solid little balls, and things made of atoms as solids, was a very workable and useful approximation. Classical physics does very well in its domain and the classical approximation is still useful. Houston put a man on the moon using simple Newtonian equations; the extra accuracy Einstein's tensor

---

1 P. W Atkins, Quanta (2nd ed.), Oxford University Press, Oxford (1991), p. 348.

2 Richard. P. Feynman, QED: The Strange Theory of Light and Matter, Princeton University Press (1985), p. 7.

equations of General Relativity would have provided was as unnecessary as telling a carpenter you want your bookshelves 10.50269288 inches apart.

Laying down the tracery on a silicon chip, however, does require such accuracy. Such happened to the early pioneers when they began resolving phenomena at the atomic level. The classical concepts turned out to be blurred-out, external approximations of a deeper, internal aspect to objective reality.

The classical concepts, so useful for billiard balls, were totally impotent to describe the atomic phenomena being explored by the pioneers at the turn of the last century.

Something had to change, and change it did, slowly. Each concept along the way had to win acceptance against powerful opposition because each concept was so counter-intuitive and bizarre.

The new physics is, indeed, so radically weird to the classical mind that it is very difficult to accept the basic concepts. As one wit put it: not only is reality stranger than you think, it is stranger than you can think. And we are stuck with the weird quantum view, which has gone from one success to the next throwing off a plethora of goodies based on electronics such as my Mac with its CD burner and DVD reader.

"Perhaps, someday, an experiment will be performed that contradicts quantum mechanics, launching physics into a new era, but it is highly unlikely that such an event would restore our classical version of reality. Remember that nobody, not even Einstein, could come up with a version of reality less strange than quantum mechanics, yet one that still explained all the existing data. If quantum mechanics is ever superseded, then it seems likely we would discover the world to be even stranger."[1]

HARD SCIENCE

Therefore, science, as it enters the new millennium, is not just multi-disciplinary; it is a discipline with something of a split personality. In the hierarchy of physics, chemistry, biochemistry, biology and evolution, the switch-over from one science system to the other is to be found somewhere between physical and biological chemistry.

These are 'hard sciences' in the sense that their concepts are precisely expressed in the universal language of mathematics.

Like all sophisticated languages, mathematics can express the same concept in many ways, from a simple outline to an elaborate filigree. This is not to say that math-speak does not have pitfalls for the unwary.

For instance, it will take a graduate to whom math-speak is a second language in which they are fluent and can think in as easily as they can in English (etc.) to instantly see exactly what the next operator is in this sequence. See if you get it instantly as well.

1	2	3	4	5	6
$-(\ )$	i^2	$0i-1$	$\exp(i\pi/2 + i\pi/2)$	$e^{i\pi/2} \times e^{i\pi/2}$	??

Did you get the answer?

It's -1. In fact, every single one of those math-speak-words is just a different way, depending on circumstance, of saying minus-one. There are others even more intimidating!

Don't worry: we will actually only call upon the simplest of these.

Incidentally, the $e^{i2\pi}$ math-speak word for $+1$ is called the exponential or transcendental operator or function. This is why I will occasionally refer to quantum math as transcendental without implying anything New Age.

---

1 Johnjoe McFadden, Quantum Evolution, W. W. Norton, NY (2000), p. 219

The only reason I mention all this is that we are shortly going to encounter an equation that looks so fearsome you might, if unable to speak math, give up in despair of ever comprehending such a monster and throw down the book. Now, when you see it, please think "it's probably just saying '2+2=4' in rococo flourishes" and do not give up.

NEWTONIAN SCIENCE

Newton is rightly considered the Father of Science as we know it. The themes he developed in classical physics have appeared throughout the scientific structure—biology might not be a branch of physics, but physics is certainly at the foundations of biology.

Of course, one is philosophically free to drop the hierarchical constraints in constructing a theory of how the world works; but the construct will be something other than science as it is practiced today. The classic historical example of this is the attempt to explain living systems by the introduction of a "vital force" in one guise or another. While there are many philosophical constructs that embrace this as an acceptable explanation, none of them are part of biology because particles, atoms and molecules can be understood without a vital force and, if electrons and quarks don't have it, neither do the atoms and molecules they comprise, nor neither do cells nor higher organisms.

So, while the biology of our era is proud of its firm foundations in the "hard" sciences (those amenable to mathematical rigor), the physics in which it is rooted is the classical physics of Darwin's day. "It is most ironic that today's perceived conjunction between physics and biology, so fervidly embraced by biology in the name of unification, so deeply entrenched in a philosophy of naive reductionism, should have come long past the time when the physical hypotheses on which it rests have been abandoned by the physicists."[1]

There is still, of course, the sense that science should be a unified structure: "How does nature encompass and mold a billion galaxies, a billion, billion stars—and also the earth, teeming with exuberant life? New insights into how nature operates come from parallel advances in particle physics and in molecular biology; advances that make it possible to examine fundamental physical and biological processes side by side. The resulting stereoscopic view deep into the past reveals a previously hidden, unifying logic in nature: its paradigm for construction."[2]

To say that pre-twentieth scientists were content with Newtonian physics, chemistry etc. is an understatement. Scientists were, almost literally, dragged kicking-and-screaming into accepting the quantum worldview because the only deity in science insisted upon it. That deity is experiment. For no matter how elegant, mathematically-sound, politically-correct etc. a theory might be, if it contradicts experiment it is crumpled and into the wastebasket.

Hopefully, by the end of the next section, you will be convinced that the new physics is truly and radically NOT the science you thought it was.

SHOCK AND CONFUSION

In the Appendix: Slit Experiment, I deal with the actual experiments that so utterly confounded the physicists of a century ago. Here, just to put things in perspective, I would like to give a feel for the shock-horror these scientists felt when they saw extraordinary experimental results that insisted that all their preciously-won-since-Newton classical theories about reality had to be thrown into the wastebasket.

To do this I will tell a short story:

In the Big House, four executions are scheduled to take place by firing squad. The squad, all armed with machine-guns, is in one room, and a post to restrain the prisoner is in another. Between the two rooms are two very large windows in the wall that can be covered with heavy steel shutters.

---

1 Robert Rosen, Life Itself, Columbia University Press, NY (1991), p. 18.
2 Edward Rubenstein, "Stages of evolution and their messengers," Scientific American (June 1989), p. 132.

On a whim, the warden decides to use the shutters to test his classical expectations. He was pretty certain as to what would happen but was prepared to test his theories against experiment:

The first experiment had both shutters closed. This 'control' lived up to expectations. The shutters over the holes stopped the bullets from reaching the prisoner and his life was spared.

The second and third setups had just one hole shuttered—first with the left open, then with the right. This experiment also "lived" up to theoretical expectations: The prisoners were each shredded by the hail of high-velocity bullets streaming through the void of the open window.

It was the fourth setup that violated all expectations. With both windows open, no bullets reached the prisoner. Not a one of the mighty hail of bullets reached the deafened and terrified prisoner. Just to be sure, the warden kept the gun firing extra time.

He found it hard believe his own eyes. Two voids stopped the bullet; while just one open window did not. Two empty openings were as effective a bullet shield as two steel shutters!!

The warden just had to know what was going on so he repeated the both-window open execution, but this time knocked holes in the walls so he could see in to watch the magic of 'nothing' stop bullets like solid steel.

Ratcheting up the warden's total stupefaction and torment, however, this time, as he was watching, the bullets behaved as expected. They poured through the holes and the prisoner was shredded very, very quickly.

The astonishment of the warden at this unexpected result and the mental gymnastics he went through trying to digest this result gives you a sense of the state of physics at the start of the twentieth century. To be true, the experiments that they had to explain did not involve bullets and criminals but to the scientists shooting electrons and atoms at detectors through slits, they might just as well have been.

This is, in essence, is what was observed in the slit experiments performed by the pioneers. Can you feel how horribly perplexed they were trying to digest such a phenomenon. The experiment violated all expectations on the most fundamental of levels.

See the appendix for an exposition of the real slit experiments.

Try yourself, using the physics you picked up in high school, to come up with a reasonable explanation for nothing acting like steel. Don't spend a long time at it, however, genius has tried and endlessly failed.

One thing was clear. There was, and is, no way to explain such a thing with the "commonsense" notions at the heart of Newtonian physics.

REVOLUTION STEP BY STEP

The path of science history from these first puzzling slit experiments to some sort of confident understanding spanned almost a century. To say that scientists were "forced" into the quantum description is not hyperbole.

The transition from the old to the new stretched over many decades and, even in these enlightened times, there is still debate about 'what it all means.' The quantum revolution was indeed a most reluctant revolution.

The 20th century was a time of transition "when the classical model of the mechanical universe became untenable and began to be modified by a patchwork of rules involving the energy quanta introduced by Planck in 1900."[1]

It was with great reluctance that scientists faced up to the implications of these changes: that their description of objective reality was horribly inadequate.

---

1 Philip Stehle, Order, Chaos, Order: The Transition from Classical to Quantum Physics, Oxford U. Press (1994), p. i.

The establishment of the current worldview of physics was not based on theoretical speculation; the current view vanquished the old not for theoretical reasons but because that ultimate arbiter of science, experiment, insisted on it.

"The quantum era had arrived but it did not bring an end to controversy. The interpretation of the new quantum kinematics was, and still is, a source of both conceptual discussion and experimental exploration of its consequences in places where it contradicts deep-rooted intuitions of physicists and others, especially for questions of physical reality and causality. So far, all the experimental tests have decided in favor of the quantum kinematics. More than that cannot be said."[1]

Scientists are compelled to accept the quantum view—sometimes with profound discomfort—because it always, without fail, agrees with experiment while the classical view, just as consistently, does not.

Chess Pixels

Even through the quantum revolution is now a century old and is perfectly described by the mathematics, translating the math into a natural language is tricky and usually controversial.

The basic firmly-established equation, for instance, that accurately describes an atom is:

$$-\frac{d^2\psi}{dx^2} = \frac{2m}{\hbar^2}(E - V(x))\psi$$

This is the fancy, baroque way of putting it. Later, we will generalize this into something much simpler. Take heart from the following equation that is almost as bad:

$$i^{200} - i^2 / (e^{\pi 2i} + e^{\pi 200i}) = e^0$$

This is actually just a fancy, and occasionally useful, way of stating that $2/2 = 1$.

There are only a limited number of solutions to this equation: The "1s orbital" of hydrogen is the simplest of these "eigenfunctions," while the largest, uranium, has electrons in the "7s orbital."

This equation of Schrödinger's is as firmly established in science as anything is. Yet, ask a quantum physicist, "What does it mean?" and you will get a variety of translations, some emphasizing uncertainty, some wave-particle duality, others non-local causality, etc.

For an excellent introduction to this equation I recommend the book, translated from the Russian, from where the following quote was lifted:

"This all implies that electrons exist in the atom not as particles but as waves, whose nature was not quite clear at first, even to Schrödinger himself. What was clear to him, however, was that whatever the nature of these electron waves, their motion must obey a wave equation. Schrödinger derived such an equation. It looks like this:

$$-\frac{d^2\psi}{dx^2} = \frac{2m}{\hbar^2}(E - V(x))\psi$$

"The equation says absolutely nothing to those who see it for the first time. It induces curiosity or even a nebulous feeling of instinctive objection (without serious grounds for the later.

"… Physicists were quick to appreciate the advantages of wave mechanics—its universality, elegance, and simplicity. Ever since they have almost abandoned [more chunky methods]."[2]

I am going to translate the math into familiar, anthropomorphic terms. And make no apologies for it; even attempt a justification for such apparent laziness towards the end.

---

1 Ibid. p. 307.

2 Ponomarev, L. I. The Quantum Dice, Institute of Physicics Publishing, Bristol (1993), p. 94.

We will need just two aspects of the revolution encoded in this intimidating hieroglyphic. They are fundamental and decidedly non-classical: quantum pixels and quantum probability.

QUANTUM PIXELS

First, the little 'h-bar' on the right side of the equation. This involves a key concept in both classical and quantum physics: that of the action, the fundamental measure of existence.

The concept of "action" and the affiliated "principle of least action," were developed by Lagrange and others in the eighteenth century as an alternative formulation of Newton's equations of motion, the basis of classical physics. The action equations are more cumbersome than Newton's in simple situations and, consequently, never caught on in classical physics—the action equation that describes the motion of a pendulum, for instance, is much more mathematically challenging than its simple equation of motion.

In complicated classical situations as well as quantum physics, however, the superiority of the action formulation is overwhelmingly apparent.

Action is a such fundamental measure of the state of systems that, in a sense, the task of science is to discover all the factors that contribute to the action of a system and the "action equation" that describes how they combine:

"Physics can be formulated with the action principle. A given body of physics is mastered if we can find a formula that empowers us to determine the action for any history... The action principle turns out to be universally applicable in physics. All physical theories established since the time of Newton may be formulated in terms of action...

"Our search for physical understanding boils down to determining one formula. When physicists dream of writing down the entire theory of the physical universe on a cocktail napkin, they mean to write down the action of the universe. [The accompanying illustration is a contemporary action equation; 's' is the total action.] It would take a lot more room to write down all the equations of motion...

"The action, in short, embodies the structure of physical reality."[1]

It is an action equation that describes the combined influence of all the many interactions on the changes in the overall history of the system. This is as true for quantum physics as it is for classical physics.

THE PATH OF LEAST RESISTANCE

Next we will deal with just what is the cause of the probability amplitude, the cause-of-the-cause-of-probability, if you will. The cause of the probability amplitude involves something that we have already discussed. A path of history, a series of interactions, generates action, the scientific measure of resistance.

The rule is that systems tend to follow the path of least resistance. At first encounter it does not seem to be the sort of thing we expect scientists to state about the world. Certainly the statement might seem more at home in many a Californian subculture.

We will now proceed, however, to explain how the statement, Systems tend to follow the path of least resistance. has a precise—i.e. mathematical—scientific description of how the world works.

In classical science, systems always follow the path of least

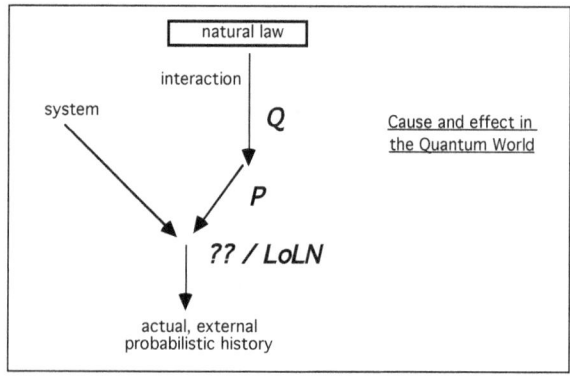

Cause and effect in the Quantum World

---

1 A. Zee, Fearful Symmetry, Macmillan, NY (1986), pp. 106 - 111.

resistance, in the new they tend to follow that path. The new science introduces with this qualification an element of choice that is quite lacking in the classical view. For a tendency to do the "right" thing implies an occasional lapse into the "wrong" thing. We will later in the discussion liken this to black-and-white external states connected with waves-of-gray internal states.

Action and interaction

In physics, then, the action is a consequence of interaction. The equation just mentioned is an example of the way a scientist would describes the overall consequence of interaction contributing action along each path of history. The simple-looking symbols are actually highly sophisticated mathematical entities—tensors, matrices, path integrals and other such esoteric shorthand—that must be invoked in order to measure the action for a particular interaction. It was this complexity that precluded the action formulation from wide acceptance in simple mechanics.

You will notice, however, that the final step in solving the equation is the grade-one step of adding the six numbers together. This final sum gives the overall action.

This refreshingly-simple step occurs because each of the terms in the action equation calculates the action generated by a particular interaction—one for the gravitational interaction, one the electromagnetic interaction, etc.—and the overall action is the simple sum of the actions generated by each interaction. Hence the simple final step in solving the equation.

The action equation for any system will be similar: the final action will be the sum of the action generated by each interaction the system is capable of—which, as noted, will depend on its coupling substructure, the subsystems it is capable of coupling with.

For each level in the scientific hierarchy—corresponding to the unique interactions found at each level in the material hierarchy—there will be a corresponding set of action equations that give the action that the system's interactions generate.

Luckily, in many instances, the situation is simpler than might be expected knowing the full hierarchical structure of systems. For instance, while the structure of atoms contains subsystems that can couple with the gluons and pions of the strong force, the equations that describe the action—and thus the history—of atoms and molecules does not include this. For none of these subsystems are involved in the interactions of atoms, they are sequestered in the nucleus. Neither the strong nor the weak force appear in the formulation that allows the path of least resistance to be calculated.

UNIVERSAL AND SPECIFIC

The application of the quantum perspective to figuring out what any system will do is conceptually simple:

a) measure the interactions along each possible path

b) measure the action generated by each interaction

c) calculate the total action along each possible path

d) Compare them all: the path of least action will be the one the system will follow.

It is the necessity of examining all possible paths that makes the action way of looking at things complicated in practice.

Natural law in the quantum world thus has three aspects:

First is the universal impulse to follow the path of least action

Second, are the specific laws that determine just how much action each particular interaction generates when it is indulged in.

Third are the laws governing the development of the internal wavefunction, and how it is connected to external space-time.

THE PRINCIPLE OF LEAST ACTION

First, the universal aspect of natural law, the PLA. The "Principle of Least Action" is one of the "givens" in our universe—it cannot be derived from any simpler principle (as if anything could be simpler!). It is sometimes referred to humorously as the Law of Cosmic Laziness; less insultingly as Cosmic Parsimony; and with dignity as the Zeroth Commandment that "Thou Shalt Not Generate Unnecessary Action." (We note here for comparison that the quantum perspective we will soon examine is not that different in that it asserts that a system will tend to follow the history of least action rather than stating that it will follow the path of least resistance..)

Without knowing anything about gravity, for instance, it is possible to "explain" why a ball falls to the ground: the gravitational interaction in staying put or moving upwards generates more action than moving straight downwards. If you know the correct equation, you can prove this. (This rule is so general that it is also the best answer to: "Why did the chicken cross the road?" as both classical and quantum science agree that it must have been because crossing the road generated the least action. Such "explanations," however, have about as much explanatory power as "God Did It" without knowing the equation—known for the interactions of a ball but not for a chicken.)

Path Integrals

Second, the specific aspect of natural law.

Quantum physics considers action of fundamental importance. "The fundamental law of quantum physics states that the probability amplitude of a given path being followed is determined by the action corresponding to that path."[1]

The probability amplitude, measured by complex little arrows, is the tendency to follow a particular history. It is consequence of the action along that particular path. This tendency has a size and a direction "pointing" in an internal dimension as already noted.

The connection between the size and direction of this internal tendency and the action along a path of history involves a somewhat sophisticated mathematical construct called a path integral. We will not go into this in any detail, just pick up on the main details.

The action itself can also be thought to have a cause—the amount of action generated by a particular interaction is a given in this universe—it is what we call a natural law. In both classical and quantum physics, the amount of action involved is determined and can be described by equations. Both classical and quantum science agree that natural law determines the action. Each interaction has a natural law that determines just what the action will be.

The basic rule in both classical and quantum science is that each interaction the system is capable of contributes to the overall action along a path of history:

The innocent-looking symbols for each interaction in the equation we looked at earlier represents "path integrals" that sum the action over each path of history: from the state the system is in to the state it could end up in.

A simple illustration of an action integral is to let the height of a curve be the action at that point in the history, so that the area under the curve represents the action of the complete path: the path integral as it is called.

The connection between the path integral over a path and the probability amplitude for that path is: The magnitude is inversely proportional to the area under the curve—the bigger the area the smaller the size of the arrow. Its angle or amplitude is derived from the perimeter of the curve. This is sophisticated math so let's leave it at that.

This is where the probability amplitude, a quantum cause-of-probability measured by little internal arrows, come from.

---

1 A. Zee, Fearful Symmetry: The Search for Beauty in Modern Physics, Macmillan, NY (1986), p. 142.

In mathematics, the connection between a curve-with-area and a magnitude-with-direction is dealt with as the relationship between vectors and their cross products.[1] We need not, fortunately, go any further into this aspect of the math.

The general principle here is that the greater the resistance (action) along a path of history, the weaker the tendency to go that way. In quantum physics, the rule is that systems <u>tend</u> to follow the path of least resistance. It is this tendency that is the internal extension of a system.

Thus, the internal "graph" of the action, from which comes the probability amplitudes, is fully determined. If we know the action equations, we can fully know the internal tendency to change state, the probability amplitudes involved.

Non-locality

The connection between action and probability amplitude raises a very interesting question: How does the electron "know" what the action along the path is going to be before it actually travels it? How does the electron at A "know" that it will generate 47 units of action going to B and only 26 going to C without actually traversing the paths first?

While this ability to "probe the future" was also implicit in the classical action equations, it could not be dealt with satisfactorily, so was ignored (though the philosophers had a field day).

Just how a system can "know" the nature of all the possible paths open to it and unerringly pick the one with the least action is currently receiving a lot of attention under the rubric of "non-local" causality.

Certain experiments on the polarization of twined photons, for instance, can only be explained if they are able to communicate with each other about their state through some subtle agency which can convey information at speeds vastly in excess of the speed of light: "Does this non-locality actually operate at the quantum level so that two photons…, although far apart from the perspective of the scientist in his laboratory, are at another level connected? Such nonlocal connections could, in fact, stretch throughout the entire universe."[2]

We are not going to delve into this non-locality as it only really has significance in when systems are not interacting and most natural systems engage in incessant, continuous interactions at all times. It will help dispel doubt, however, when we get to discuss cells and fields of interaction that, at least, envelop a cell even though the external components exist on a much smaller, molecular level.

We can speculate, however, that this phenomena is explainable with probability amplitudes just because the little arrow is not pointing in external space-time, it is pointing in an internal dimension that is not space-time constrained. The internal extension is not constrained to the spatial extension of the system. Then internal development and change—described by the addition, multiplication and collapse of complex numbers—are communicated throughout what we will call the wavefunction independent of time.

This is consonant with experiment and theory. A single electron has a wavefunction throughout the universe—the only reason we can even think of a single electron as having a location—and single electrons have been trapped in quantum wells—is that its wavefunction is effectively zero throughout most of the universe except in the quantum orbital—it cannot "teleport" itself out of such a deep well.

Wavefunctions are not spatially limited by the speed of light in the two-open slit experiment: as the electron is leaving the source, its wavefunction is already interfering with itself at the detector.

Again though, this is not all that important as interaction interrupts constantly. With each collapse in the wavefunction—a probable event actually happens—a new wavefunction is established by the system in this new state and history progresses.

---

1 Hoffmann, B., About Vectors, Dover Publications, NY (1975), p. 101.

2 F. David Peat, Einstein's Moon: Bell's Theorem and the Curious Quest for Quantum Reality, Contemporary Books, Chicago, IL (1990), p. 146.

So we can expect that the influence of the internal aspect of systems will not be directly limited by time and space—constraints like the speed of light that so limit external things, etc.—but indirectly by sequential collapse in interaction and the passage of time.

Development of the wavefunction

The way the internal state of a system changes over time is usually called the "development of the wavefunction."

The Natural Laws in the new physics are the principles which govern the development of the wavefunction. For the elementary particles, the natural law is described by the Schrödinger Wave Equation. This equation describes exactly how the wavefunction changes and develops over time. "The most important lesson to be learned from Schrödinger's equation is that the time evolution of a quantum system is continuous and deterministic."[1]

Solving this equation for the electron, for example, enables one to calculate exactly what the wavefunction will be in the future. Unfortunately, this equation is fiendishly difficult to solve with current mathematical techniques so that only relatively simple situations, such as an electron and proton interacting to form a hydrogen atom, are fully solvable.

If this rule-of-internal-law holds for fundamental particles, we can expect it to hold for all systems composed of them. Each level in the material hierarchy will have its own natural law running the internal aspect of things by determining the amount of resistance generated by interaction—by systems externally sharing their subsystems.

Our current scientific understanding of the probability amplitude does not stretch much beyond simple molecules. With appropriate simplifications, perhaps as far as macromolecules. It most certainly, however, does not reach up into the realms of biology, genetics and evolutionary theorizing. The rumblings in the basement have yet to shake the battlements. But, even though the details of the action equation at these sophisticated levels are not known, we can expect that they are there.

It should be noted that, while all physics and some basic chemistry is formulated in terms of least action, the "laws" that biochemists, biologists and evolutionists (not to mention the soft sciences) are rarely, if ever, formulated as a Principle of Least Resistance. All physicists can comment is that they must then either be wrong or mere approximations of a more subtle level of natural law which can be so formulated.

We now have a sequence of cause and effect: probability from probability amplitude, probability amplitude from integrated action to the action being determined by natural laws.

We will not get into the detailed math description at this point. For our purposes, it is sufficient to think of the principle of least action operating on each path's action to associate it with a probability amplitude for that path.

PLA (action along paths) = p@a for path 1

p@a for path 2, etc.

We can add this to our diagram of quantum concepts and their basic mathematical description.

Yes, I know it's looking complicated. But once we have all the details of each conceptual step well-established we will do some drastic simplification and condensation of the picture. But for now, the detail.

We are now ready to tackle the other branch in the sequence of quantum cause and effect, that sitting off to the left in the diagram, the "solid matter" that gets to do the choosing.

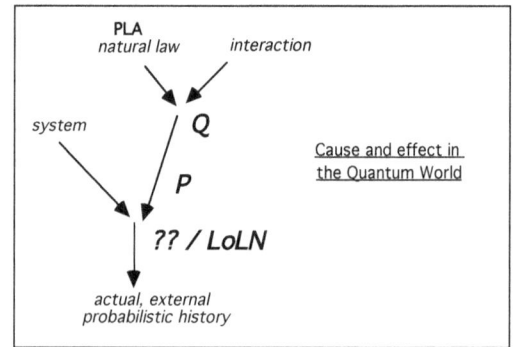

Bits of Action

In classical physics, energy-in-time and the amount of action was, and still is by non-scientists, considered to be continuous. This turned out

---

1 J. Baggott, The Meaning of Quantum Theory, Oxford (1992), p. 72.

to be incorrect.

This involved a question first tackled extensively by the philosophers of Classical Greece: How finely can you divide the things in objective reality?

There are basically only two ways to go: you can go on dividing forever, or you cannot go on dividing forever. In math-speak: reality is continuous—the forever case—or reality is discrete, its 'atomic'—you have to stop somewhere down there.

Continuous or Discrete

Take water, for example. It would seem that, no matter how small a drop of water one could imagine, it would still be water. If you cut an apple in half it is still apple, as is a tiny sliver. There is a seeming continuity here.

Take my son, however, and cut him in two and things are quite different. Half a teenager is no longer a person. Things here are discrete.

A very simple example is the difference between climbing a slope and climbing a stair.

The slope is a continuous set of states: It makes sense to say I am 1.0 foot up the hill, now I am 1.5 feet up the hill, now, getting slower, I am exactly 1.8765 feet up the hill, etc.

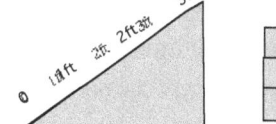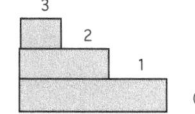

The steps, however, are a discrete set of states. It makes sense to say I am on step 1, now I am going from step 1 to step 2, now I am on step 2. It does not make sense to say that I am on step 1.5, let alone step 1.8765.

We can expand this into a 2-D illustration with multiple possible paths.

Continuous graph: A simple 2-D illustration of continuity is that of graph paper. I put a chess piece—the mighty Queen—down on the paper. The readout of a laser ultra-GPS ruler tells me that its center is, to an accuracy of four decimal places, 1.000 inches north, 1.00 inches east of the origin. With an ultra robot arm, it is no problem to shift the Queen by a circuitous and continuous route to exactly 1.876 inches north and 2.123 inches east, and then to 2.8888 N and 3.7171 or anywhere else for that matter. Moreover, it covers an innumerable number of points on its journey, as there is an infinite 2-D continuum of places to go.

Discrete chess: My favorite illustration of 2-D discrete steps is that of chess. A common opening move is to boldly claim the center by moving the pawn on 'state' e2 a 2-square jump to e4. Much less common would be the more timid 1-step move to e3 (the ghost). In standard notation this is:

1. e2 – e4 … (or 1. e2 – e3 …)

Either move is a single step in 2-D space—it is discrete. If I were to tell my computer's chess program to open with

1. e2 – e3.125

it will beep, sulk and most probably crash. For there is no such in-between state; only the before-state of e2 and the final-state e4 are relevant.

This is the essence of the discrete jump. The time it took to move is irrelevant, as is the path taken by the piece (it might have been dropped on the floor), as is just where in the final square it ends up resting—"more to the top and over just a little" is not meaningful in chess.

This is actually a good example of a quantum jump. 'Quantum' comes for the Greek for "a little bit." An electron, for instance, can be in one quantum state or another but not anywhere in between them, and the jump takes no heed of time-and-space constraints (AKA teleportation). And, like the knight move in chess, nothing can block it in.

Even the notation for the quantum electron states in atoms has similarities to chess notation. This, for example, is how scientists describe the quantum jump responsible for the bilious, if illuminating, yellow of sodium streetlights—as electrons by the gazillions monotonously making the quantum jumps:

3s − 3p 3p − 3s

ad infinitum, back and forth. Much, much less common, is the quantum jump 4s − 3s that is a lovely violet color. Just like in chess, however, asking a quantum physicist the color of the quantum jump:

3s − 3.125 p

will just prompt disparaging mutters such as "modern education!" and "hopeless, hopeless!"

Classical science soon found many of the things that they had assumed were continuous were actually discrete.

Cutting an apple has limits: when you get down to a single cell it is still apple, but dividing that cell in half is just like cutting my son in two—the two halves are no longer apple, they are cell debris. A drop of water can be divided much further until reaching a single molecule—cutting that no longer gives you water but atoms of hydrogen and oxygen. Matter was discrete.

But other things, including time, space, energy, light, spin, etc., all seemed to be continuous in nature, and were decidedly so in classical physics.

This turned out to be wrong. All of them turned out to be quite discrete and chess-like. The reason for the apparent continuity was resolution: a vast chessboard seen at a distance will not be checkered at all, it will just look an even gray and apparently continuous.

Physics describes pixels in the real world that are decidedly small by our standards, which is why, from our perspective, the jerkiness, is not apparent.

The 'pixels' of reality are so tiny, however, that the classical approximations of continuity—much the easier to describe in mathematics—still serve us very well. It is only at 'natural' resolutions that the quantized 'squares' of reality—of existence itself as action—become apparent.

So, from a distance, my PowerBook's screen seems continuous; the curves to the black letters seem crisp and sharp. Up real close, however, I can see square pixels and a bad case of the jaggies.

In a similar fashion, when scientists developed sophisticated devices to zoom in on reality, they found that everything came in pixels, nothing was continuous. Space, time, energy, charge, gravity, spin—you name it, they all came in pixels of a certain size.

Luckily for our sanity, reality is ultra-ultra-high resolution; the pixels are so, so tiny that we do not consider existence to be jerky.

Back to the Schrödinger equation of the atom and the little 'h' that appears on the right-hand side as its inverse squared. This conversion factor, h, for the action into pixels is known as Planck's Constant and, to a high accuracy, it's:

0.000000000000000000000000000000000000

00000000000000000000000033511346 lbs secs

Tiny indeed, which is why reality seems so smooth. We just can't see the jaggies no matter how hard we concentrate. The famous uncertainty relation is related to this 'pixel' aspect of quantum unexpectedness.

This pixelation of existence—the real world about us—has odd consequences.

The tick of time is very, very small. Even quite big bits of matter when multiplied by such a small number can still remain under the pixel of existence limit. So there is no reason why the bits of matter should not appear for a few q-ticks of time.

If a speck of matter were to appear out of nothing for just a q-tick or two before dematerializing then it would not "officially" exist. There is then no reason then, in a pixelated reality why specks of matter should not appear out of nothing and then disappear back to nothing. No "real pixel" of existence/action is created so that classical, approximate laws—such as the conservation of mass/energy—are not "really" violated.

We can define a virtual crime analogy in the penal code: stealing a wallet and then returning it unaltered to its rightful owner before the theft is noticed is, technically, a violation of the laws protecting private property. But it's not a 'real' crime, a pixel of law-enforcement response, so to speak, is not generated, it's a virtual crime. The bigger the item you steal, the less time you have before returning it and avoiding detection.

Now one of the simple rules of quantum physics is called the totalitarian principle:

That which is not forbidden is compulsory.

If there is nothing forbidding a speck of matter from popping into the universe for a q-tick or so then it is compelled to do so. And it does so. Naturally you cannot directly detect such fluctuations of the vacuum into matter and back—they would have to officially exist for that—but such "virtual" particles have been experimentally confirmed by their indirect influence on real things such as electrons in atoms. Note that the smaller the lump the longer it can hang around before generating a bit of existence and thus violating the conservation of mass and energy. For instance, the empty vacuum is actually a froth of virtual electrons and their antimatter counterparts (and all sorts of other things that are not forbidden).

Like virtual particles, indirect effects of virtual pick-pocketing might be observed. In our analogy, crossing Grand Central Station at rush hour with intense virtual pick-pocketing along the way might cause an otherwise-unexplainable fraying of the wallet pocket. In our society, Luckily, virtual pick-pocketing is not compulsory, or at least, not that I've noticed.

LIMIT TO KNOWLEDGE

In classical science there was the possibility of ever increasing accuracy in the measurement and knowledge of paired aspects of things such as position and momentum or duration and energy

So, for instance, the ratio π in mathematics has been calculated to an accuracy of billions of decimal places. In classical theory there was no reason to think that, given sufficient technical advance, position and velocity or duration and energy of things could also be accurate to an ever-increasing number of decimal places.

This, again, turned out to be incorrect, there is a limit to what can be known. This is the Uncertainty Principle that gets a lot of attention in most books about quantum science.

The Uncertainty Principle limits how accurately such paired attributes can be known: the better you know the one the less you can be certain about the other. Measure with accuracy the momentum of the pea-sized electron in a Yankee Stadium-scaled atom—which can easily be done—and its position could be anywhere within the stadium.

The combined precision of known momentum and position is measured in units of Planck's Constant. So this limit-to-knowledge is actually just another way of looking at the pixels of existence—you can only pin down things down to a single pixel; fractions of this are not 'real' and are thus unknowable.

EINSTEIN'S NOBEL INSIGHT

In all areas of the new physics, it became clear that the world obeyed the quantum, chess-like rules of the discrete, not graph-like continuous ones. We shall give a brief example that involved Einstein's prize-winning contribution to the Quantum Revolution. (Not for relativity.)

It involved the nature of energy. Is it continuous or discrete? It certainly seems to be so: classical physics was quite capable of measuring changes in energy to an accuracy of a dozen or so decimal places of a watt.

In classical physics the assumption, rarely noticed, was that energy was continuous, you could spread it out as thinly as you wished. We are all familiar with the fact that light intensity falls off with distance: the glaring headlights that cause temporary blindness just 10 yards away are hardly noticeable 10 miles away on a Great Plains interstate. What about at the distance of the moon? The next galaxy? The ends of the universe? If energy is continuous, with sufficiently-sensitive ultra-instruments, we should be able to follow the intensity as it gets closer and closer to zero without it ever reaching zero exactly.

Einstein won his Nobel for showing this to be incorrect: energy actually comes in distinct, and decidedly discrete packets.

He won by being the first to successfully explain a phenomenon that is often used these days in automatic door openers—the photoelectric effect.

The electrons in some metals are very loosely held and float freely, hardly held at all by any atom (a fact that underlies electricity and the shiny look of all metals). Just a little light suffices to kick huge numbers of electrons from some metals.

Einstein came up with the prize-winning explanation of the following, quite unexpected, result of experimenting with this photoelectric effect:

A metal is exposed to red or blue light and the number of electrons kicked out is measured. One light source is more intense than the other:

Red light: a 1,000,000,000-watt searchlight.

Blue light: a 1-watt Christmas tree decoration.

The classical expectation, if light energy is continuous, is that the billion-watt red light will kick out a lot more electrons than the feeble one-watt blue light.

Unfortunately, classical expectations were exactly wrong; the result that Einstein successfully explained was the ineffectiveness of intense red light compared to the effectiveness of the blue:

<u>Red</u> <u>Blue</u>

0 1,000,000,000,000

Einstein received his Nobel for coming up with the following combination of ideas already 'in the air' as his genius flourished:

1. Energy is discrete: it comes in distinct, particle-like, packets called photons—bits-of-light so to speak. It is utter nonsense to speak of half-a-photon of blue light, let alone 0.0123 of a red one. This is the nature of the quantum, chess-like jumps that characterize objective reality.

2. A single packet of blue light has more energy than a photon of red light.

3. There is a quantum jump between the bound and the free state with no in-between states. Chess-like e2-e3 behavior again.

This brilliantly explains everything as we can see with this simple diagram.

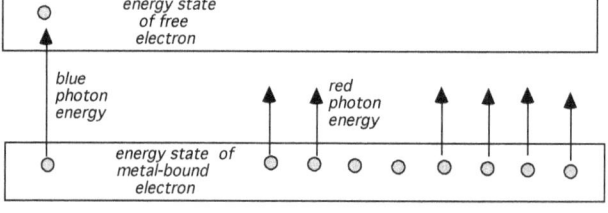

. The red light is not energetic enough to lift the bound-state electron into the free-state (the chance of two reds getting absorbed at exactly the same time before the red energy is reflected or turned into heat is infinitesimal.) Just one blue photon is quite sufficient, however, to jump an electron into the excited, freed state. And even a 1-watt light is gazillions of photons (so they can trigger the opening of your garage door, perhaps).

Simple; once some genius has figured it out first.

Pixels of spacetime

Even the continuous map has its limitations, it turns out, as on a really, really tiny scale even space-time itself is discrete with pixels we can call q-ticks and q-spans.

The quantum q-tick of time (officially the Planck Time) is about:

1/1,000…(40 zeros)…000 of a second.

It is as much nonsense to state that something takes only 1/2 a q-tick or 3.12345 q-ticks as a pawn in chess moving to e3.5.

The quantum q-span (officially the Planck Length) of space is about:

1/1,000…(36 zeros)…000 of a meter.

Our universe is about 15 billion years old and stretches about 15 billion light-years in each direction. I can figure out how many pixels there are on my Mac monitor by multiplying its dimensions X and Y:

number across x number down = 1280 x 854

= 1,093,120 pixels

In the same way, we could calculate the total number of spacetime pixels in our historical universe by multiplying T, X, Y and Z:

number q-ticks in 10 billion years   x

(number of q-spans in 10 billion light years)^3

You can go figure it if you want but, be warned, it's a really, really big number.

Classical science has no concepts that can deal with discrete pixels of energy, space and time.

As is apparent, it is this chess-like stepwise 'quantum' nature that has given its name to the entire revolution.

The quantum world generates the appearance of a reality described by classical science. Just as well, a world with big pixels would be like looking at a really old movie. And people would just disappear in one place and appear in another. Travel would be sickening as the scenery flicks from one scene to another and we traverse the stepped landscape like those little moles popping up in the fairground game of hit the mole.

The Law of Large Numbers

Mathematicians chanced upon the Law of Large Numbers (LoLN) first when trying to describe games of chance, gambling, gaming, etc. Basically, the LoLN is the common-sense-notion that, in the long run, events reflect their probability.

In classical physics, this law is useful in statistics but is not considered to be all that fundamental. Some classical examples of the LoLN at work:.

If the chance of tossing a head with a fair coin is 50% then the chance of throwing three heads in a row is 12.5%, ten-in-a row is about 1/2,000, twenty-five-in a row is about 100 million to one—odds regularly encountered in state lotteries. The chance of a hundred-heads in a row is about one in a trillion-trillion-trillion—essentially, but not exactly, zero.

Its like the rationale I use to justify spending a $1 on a MegaMillions ticket for a $100,000,000.00 jackpot. The difference between having exactly zero chance of having $100 million dollars—my non-ticket-owning prospects—and the chance of having $100 million dollars not being exactly zero—the ticketed state—is worth every penny. Buying more than one ticket, let alone a fistful, however, increases this minuscule probability by so very little that I rarely buy more than one.

The law of large numbers guarantees that, given that sufficient people play, someone, somewhere, is going to throw "twenty-five heads in a row" and make that quantum jump to mega-wealth. Not all the lotteries in all of history would have a winner for the hundred-in-a-row, however. The number of attempts is just way too small. Atoms such as uranium, in their superabundance, exhibit detectable radioactivity even though the probability for each atom is so, so tiny.

On the other hand, the law of large numbers guarantees that, if you toss a coin a sufficiently-ridiculous number of times you will, eventually, get a hundred heads in row. If you buy a large enough number of lottery tickets the LoLN guarantees you will win the jackpot. (Beware the fine print to the innocuous-enough phrase, a large enough number. This number is large enough to make even the Pentagon's annual budget seem a pittance.

This aspect of the LoLN is just that which is not forbidden is compulsory in another guise: Either something has a probability of exactly zero to an infinite number of decimal places or its probability is not exactly zero even if a gazillion decimal places are involved.

There is one more aspect of the law of large numbers, one that we will often see in action.

Compare the probability of throwing a head with the actual number of heads we get when we throw real dice. If we just throw a few coins we might see a significant deviation from the probability: a not-unlikely four heads in a row is 100% heads, nothing like the probability of 50%. If we throw 1,000 tosses we will probably end up with something close to 50% e.g. 561 heads: 439 which is a ratio of 56% which is much more like it. Reality is reflecting probability with more accuracy the larger the numbers get. How about a trillion tosses: the heads will be 50% to many places of decimals.

It is this aspect of the law of large numbers, and this alone, that prevents the gazillion atoms of air in this room from all ever being in the other half of the room at the same time, leaving me gasping in my half of the room in a total vacuum. For, just like a fair coin, air molecules move so fast and freely that every microsecond they make a choice: this side of the room or that side of the room with essentially 50-50 probability. The numbers involved are so huge that its 50% in each half of the room to dozens of decimal places.

The basic rule is: the more you do it, the more reality will reflect the probabilities. The law of large numbers is also the reason why gambling, sorry, gaming casinos do not quail when a 'whale' wins a few million dollars. It has been taken into account. For, the in-built house advantage—and it varies from about 1% for blackjack to a usurious 30% in keno—is guaranteed by the LoLN to be the return on the gross. The only caveat is that there be a large-enough number of gamblers with disposable wealth flowing in the doors, which does not seem to be a problem. Over the long run, the house is guaranteed to make its 1% on blackjack and 30% on keno. The probability gets 'fleshed out,' so to speak, given large enough numbers for the LoLN to kick in.

In classical physics, the law of LN is useful but not fundamental. In quantum physics, however, it plays an essential role.

For instance, it is the LoLN in the new physics that is responsible for the classical—and now relegated to a 'useful approximation'—concept of 'solid matter.' It is how, as we shall see, a light, pea-sized electron hovering about a grapefruit-sized, massive proton can seem to be, and behaves as if it were, a Yankee Stadium-sized 'solid' atom as the electron teleports about within the bounds of the atom

Hierarchy

Of course, classical and quantum science do not disagree about everything; they are in complete agreement about the external aspects of matter.

One basic agreement is that all things are composed of simpler things. A basic tabulation of the hierarchy of non-living systems and their constituent subsystems is remarkably compact:

Systems	Subsystems
Molecules	Atoms
Atoms	Nuclei and electrons (cool)
Atomic nuclei	Nucleons
Nucleons	Quarks
Quarks & electrons	?

As far as the non-living systems are concerned, this taxonomy of systems is complete except for one not-so-minor detail—it seems to only embrace about 10% of the universe: the visible part. It has been established, through astronomical observation and cosmological theory, that the other 90% of the universe is made up of "dark matter" which is not visible (which accounts for why no one noticed it until recently).

This must surely be the coup de grace to the historical trend that—not content with moving the earth and its inhabitants from being at the very center of the universe to being a minor planet about a star among 100 billion others in our galaxy which is just one of 100 billion others visible to the Hubble—has relegated the grandeur of our multi-galactic visible universe to being a minor component of a much larger reality which, as yet, we are only vaguely aware of.

While there is no consensus as to what this dark matter actually is, most theories limit its possibilities to some sort of exotic elementary particle or, less convincingly, to various kinds of non-luminous "regular" matter.[1]

The previous was written in the 1990s. Things are now even stranger. I was informed at the proofing stage of this manuscript that, "I believe the current tally of the universe is 70% dark energy [Einstein's cosmological constant in reverse with a vengeance], 25% dark matter and 5% normal matter."[2]

Plus one hundred billion photons and neutrinos for each atom, I might add.

In either case, the dark matter is either made of quarks and electrons or can be lumped in with them as "fundamental systems whose structure is currently unknown."

FUNDAMENTAL PARTICLES

While it is true that the structure of the fundamental particles is currently not fully known, modern physics does have some idea of what subsystems are to be found inside an electron or quark system. As this is required for the discussion of interaction in the next chapter, we will take a moment to explore this frontier of late-twentieth century physics.

These systems on the bottom rung of the hierarchy—the indivisible "atoms" of our age—do not seem to have any inner structure when probed with high-energy collider "microscopes" and, to some, are at rock bottom in the material hierarchy and exceptions to the principle of "systems of interacting subsystems." Their appearance as featureless points, however, is more plausibly explained as a limitation on current experimental methods, which can only "see" structures on the scale of 10^{-16} meter. There is speculation that there is inner structure on a much smaller scale:

Just as "the proton… [is] formed from three quarks… the electron … [could be] formed from three very heavy new subquarks, all tightly bound …. Might not a subquark then be composed of three even heavier sub-subquarks or sub2-quarks? Extrapolation almost forces one to postulate a progression of new subX-quarks, smaller and smaller,… held together by new, stronger and stronger forces…."[3]

However: "Following this frenzy [of resolving bare quarks] we seem to have hit the experimental basement. Even if you turn up the magnification by an additional factor of 1,000—as you can at Fermilab or CERN—there appears to be no more layers of matter, no further strata. Bedrock down until '?'"[4]

SUPERSTRINGS

Another line of speculation on the structure of the elementary particles is Superstring Theory, which theorizes that what we call an electron is a distortion in a space-time continuum with 26 or so exotic dimensions (or is it just 10, I forget) in addition to the four familiar dimensions of space and time.[5] A theory that abandons the concept of classical physics that space-

---

1 M. Longair, "The New Astrophysics," in The New Physics, ed. Paul Davies, Cambridge University Press, New York (1989), p. 199.

2 Dr. David Burton, U. of Bridgeport, CT. Personal communication, May 2005.

3 Hans Dehmelt, "Experiments on the Structure of an Individual Elementary Particle," Science 247 (1990), p. 544.

4 Timothy Paul Smith, Hidden Worlds p. 150.

5 F. David Peat, Superstrings and the Theory of Everything, Contemporary Books, Chicago, IL (1988), p. 97.

and-time are a featureless stage upon which matter particles move. In this scenario, the subsystems of the "elementary" particles are vibrational modes and topological constructs in this poly-dimensional stuff.

We are not aware of these extra-dimensional extensions, the theory explains, because they exist on a scale on the order of 10^{-34} meter, extraordinarily small even on the subatomic scale. Properties of particles, such as electric charge, are the result of particular kinds of twists and deformations. The obvious "next question" as to what dimensions are made of—be they the space, time or the exotic variety—has yet to be answered convincingly.

One of the compelling reasons why Superstring Theory is being taken seriously—even though it seems to violate Occam's Razor of no unnecessary hypothesis—is that it is consistent with Einstein's work, which established that the familiar dimensions of space-time have a topological structure, a curvature that is perceived as the phenomenon of gravity. Our universe is basically flat—with local curvatures, such as that about the sun, being pretty mild—except for exceptions in the vicinity of neutron stars, black holes, etc. This is why we can, on a dark night with a good telescope, see for quintillions upon quintillions of miles in all directions.

The exotic dimensions, in this plausible theory, also have a curvature to them, though in their case it is anything but mild. A super-gravity puckers them up to "universes" on a scale so small that, to them, an electron is as the earth is to a grain of sand. In the Big Bang origin of the universe, all the dimensions started off with this tremendous curvature and just why the exotic dimensions remained crumpled up while the familiar three vastly expanded, will presumably emerge as some theory gets a better grip on what gravity is.

THE VACUUM

What is known about the inner structure of the "fundamental" particles is based on the totally non-common-sense perspective that quantum physics has found to be the best description of what "nothing" is—the nature of the vacuum. The common-sense view is that a vacuum—a volume of space empty of all particles—is just nothing, the empty stage upon which particles move about.

The classical view, similar to common sense, is that "something" and "nothing" could not be more different. The quantum view, however, is that they are actually so similar that "nothing" easily turns into "something," and just as easily turns back to "nothing" again.

Classical physics viewed reality as particles of material existing in the nothingness of the vacuum. For example, an electron (clearly a "something") was considered as moving in the absolute emptiness of space (the epitome of "nothing"). In the classical framework, the vacuum is essentially different from matter; it's what you are left with when all matter is removed.

The quantum-mechanical description of a particle such as an electron, however, is so similar to the quantum-mechanical description of the "empty" underlying space/time in which the particle moves that the vacuum has a distinct tendency (technically, a "probability amplitude") to change into particles. Such materialization is always in the form of a particle-pair, a particle and its anti-particle such as an electron and positron (anti-electron). As a positron is just as much a material particle as is the electron, we are quite correct to view this as the creation of matter out of nothing.

As might be expected at this point, when a particle and an antiparticle recombine, we end up with no particle; we are back to the vacuum.

The reason why such materialization always involves particle-pairs is geometrically reasonable in superstring theory: the undeformed vacuum when "twisted" will produce complementary curves, such as "left and right" handed, which we label "particle and antiparticle." When a right- and left-handed twist gets together, you end up with no twist at all. In non-string theories, the rationale for pair formation involves conservation laws such as charge.

This is as simple as the sequence 0=1–1=0.

As twisting space-time takes energy, one constraint on this materialization is that it cannot create a quantum or unit of existence (as will be described in the section on Planck's Constant and the "action"). In the absence of an energy source, this restriction limits the time such a particle-pair can exist to extremely small fragments of a second before reverting back to nothing again.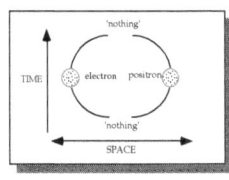

This ephemeral existence—much too brief for our current techniques to directly observe—earns such particle-pairs the designation, "virtual particles." As far as quantum physics is concerned, however, these virtual particles are the same as the less-ephemeral "normal" particles; just not very long lived.

In the quantum view, "nothing" is so similar to "something" that they easily interconvert. This strange creation of "matter" out of "no matter" is clearly at odds with the 19th-century view in which matter most certainly would never appear out of nowhere. Yet, the quantum view is thoroughly supported by experimental evidence and is the only explanation of phenomena such as the Casimir Effect[1] in which the virtual particles exert a measurable pressure. As we shall see, this "something-out-of-nothing" phenomenon is central to the current explanations for the interaction of subatomic particles. As might be expected, the creation of matter out of nothing is thought to have played a central role in the origin of matter in the Big Bang, a discussion of which can be found in Appendix: Origin of Matter.

Contradicting common sense, the vacuum is a bad example of nothing; it has a structure, often described as a "quantum foam," made of these ephemeral, virtual particle-pairs. This vacuum structure is integral to the structure of regular particles and is included in the subsystems making up fundamental particles.

---

1 P. Knight, "Quantum Optics," in The New Physics, ed. Paul Davies, Cambridge University Press, New York (1989), p. 297.

Quantum Probability

The other thing we note in the equation is the Greek letter psi, that odd looking 'Ψ' that appears on both sides.

$$-\frac{d^2\psi}{dx^2} = \frac{2m}{\hbar^2}(E - V(x))\psi$$

This mathematically represents an aspect of reality that goes by many names in natural language: the wavefunction, the 'final probability amplitude,' the internal cause of quantum probability, etc.

The wavefunction is not something that can be measured with regular numbers. Only complex numbers will do. Belying their name, these numbers are actually quite simple to understand and I have included a primer in the appendix for those unfamiliar with their delightful properties.

The crowning triumph for the developers of the new physics was the establishment of a highly accurate mathematical description of the electron and photon using such complex numbers to measure the objective extension of things in an internal space.

The techniques of QED are technically known as "Feynman diagrams," but are often called "adding little arrows"[1] because complex numbers are usually diagrammed that way.

In the new physics, such "little arrows" describe the probability amplitude. Their combining by addition and multiplication describes the wavefunction while their square describes the transition from the internal world of the probability amplitude to the external world of probability (which is always real and positive.)

I cannot resist a "what does it all mean; how to translate the math into English" comment here: If the probability amplitude is an arrow measuring an important aspect of objective reality, just exactly where are they pointing? The physicists call wherever it is they are pointing an extension in an "internal" space to distinguish it from the external space-time extensions so well described by classical physics.

Simple Rules

QED was the first theory to use probability amplitudes to deal successfully with the experimental challenges to the classical worldview. While QED can get mathematically forbidding after just a few pages, at heart, it is remarkably simple—it is basically the iteration of just three rules. Using these simple rules, QED can calculate the probability of all the possible histories—or sequences of events—that could happen to a system.

The simple rules[2] are:

Rule 1. Add complex numbers: If the history can occur by many different paths, <u>add</u> the probability amplitude for each path to get the final probability amplitude for the history.

---

1 Richard P. Feynman, QED: The Strange Theory of Light and Matter, Princeton University Press (1985), p. 27.

2 J-M. Lévy-Leblond & F. Balibar, Quantics: Rudiments of Quantum Physics, (trans.) North-Holland, Amsterdam (1990), p. 173.

Rule 2. Multiply complex numbers: If the event occurs by a series of sequential steps, <u>multiply</u> the probability amplitude for each step to get the final probability amplitude for the event.

Do this for every possible thing that could happen. This final combination of little arrows is the wavefunction, the psi in the Schrödinger equation.

The final step and the simplest:

Rule 3. Square the final set of complex numbers, the wavefunction: Transforms the internal probability amplitude into an external probability. In quantum science, the probability of an event is the absolute square of the probability amplitude for the event.

All three involve complex numbers and their combinatorial properties—a real number only pops up at the very end with step three. This is basically how science describes the quantum cause-of-probability.

TOO MANY COOKS

This is what must be the best illustration of the difference between a probability and probability amplitude:

"An amplitude is less definite than a probability: it is a sort of tendency, but as we saw it is a tendency that can help or hinder, it may be positive or negative. If you had several cooks in the kitchen each with a certain probability of making soup then the more cooks there were the more soup you would expect to get ... [This situation could be measured by real numbers such that if the probability for cook 1 to make soup is the real number P, and that for cook 2 is also P, then probability of soup for dinner $P + P = 2P$.] However, as everybody knows cooks are not like this. Cooks should rather be said to have a [probability] amplitude to make soup. Even two cooks may interfere."[1]

Real numbers are incapable of describing this, but complex numbers can.

If the tendencies of the cooks are in opposite directions, one is the negative of the other, their sum is zero and the probability of soup for dinner is zero, not twice as likely as might be expected if one didn't know the contradictory attitude of some cooks and how their temperaments combine in complex ways.

$(+1p^2) = 1P = (-1p)^2$

$(+1p + -1p)^2 = 0$

This is so much nicer than my "weird execution" scenario, but it does not convey the genuine mystification of classical science when confronted with the phenomena earlier.

It is just this behavior of complex numbers compared to real ones that encapsulates the difference between quantum and classical descriptions.

MANDELBROT SET

Underlying this description of wavefunction form are complex numbers and the way they combine with each other. The "shape" to the slit experiment is the wave-like way complex numbers interfere with each other. A short diversion is in order here to note that complex numbers seem to have an innate tendency to create interesting and sophisticated forms.

We have already noted that everything about the internal extension can be described by the mathematics of adding and multiplying complex numbers. A well-popularized development in mathematics during the last few decades specifically deals with the form-creating properties of complex numbers. Perhaps the most famous of these developments involves what has been called "the most complex object in mathematics"—the Mandelbrot Set—which has such a striking and complex form that is has been featured on the cover of Scientific American and graced countless books, computer screens and dorm walls.

---

[1] A. G. Cairns-Smith, Evolving the Mind: on the nature of matter and the origin of consciousness, Cambridge (1996), p. 241.

The Mandelbrot Set is created by massively iterating the squaring and adding of complex numbers. This is an example of a simple operator—you drop in a complex number and out pops a new number. Drop that one in to the operator and out comes another number. Repeat ad infinitum. It is not immediately obvious that this simple process could have anything to do with form but it was discovered that, when a complex number is run through this iteration, it can be classified by the two basic things that can happen:

1. The number gets larger and larger and moves off, with increasing rapidity, towards infinity. Numbers which behave like this are not in the Mandelbrot Set. The speed with which they race off to infinity is often used in coloring the set.

2. The number does not get larger and larger. These numbers belong to the Mandelbrot Set (ignoring the complication of connected sets and the disconnected sets, or dusts, at the boundary.) The number does change each time it is processed by the operator but it remains within certain limits—called Julia sets—a bounded set that gets filled in as the numbers jump around. This is reminiscent of subsystems filling in a wavefunction. The boundaries to these Julia sets have all sorts of delightful forms to them.

The fascination with the Mandelbrot Set, which can be considered the catalog of all Julia sets, is that the boundary between the numbers in the set and the numbers not in the set has an abundance of complex forms to it. In the following magnifications of the Mandelbrot Set, the coordinates of numbers in the set are colored black, those not in the set are white. Each square is a successive enlargement of the previous one.[1]

The horizontal (real) axis of the initial view is from -1.5 to $+0.5$, the vertical (imaginary) axis from $-i$ to $+i$. The final view is centered on the complex number $0.08378791+0.65584142i$ at a magnification of 34 million relative to the first view—the original is now solar system-sized and there is no end in sight!

It has been proved that the depth of form is infinite, no matter how much you magnify the set—to billions or trillions of decimal places—the forms keep on emerging. Most astonishing is the emergence of miniature Mandelbrot sets at great magnification—and the whole process repeats itself all over again if not exactly. A poet might say that very nature of complex numbers seems to be pregnant with an abundance of form. End of digression and back to the real world and the way that wavefunctions—described by the addition and multiplication of complex numbers—determine the varied forms of natural systems.

This is what the 4f orbitals look like: the different shade lobes are the wavefunction going this way and that way internally [2]. Intricate shapes and fine detail, a consequence of the internal blending of waves, is something to be expected, according to quantum science.

Internal Amplification

This transition from internal to external is referred to as "the collapse of the wavefunction" and is still a topic of lively, and sometimes bitter, debate as to "what it means" a century after it was discovered.

Yet, the math itself is so simple. Every quantum calculation always ends with the final step of squaring—transforming an internal probability amplitude, unobservable and measured with complex numbers, into an external probable history measured by observers with regular, real numbers.

---

1 Created with Super MANDELZOOM 1.06 on a Mac Plus computer.
2 www.shef.ac.uk/chemistry/orbitron/AOs/4f/index

This transformation is like squaring the familiar inch. An inch is just an ultra-thin line. Square it, however, and it turns into a postage stamp square inch, something quite different from two thin lines by themselves.

Everything interesting happens on the level of the internal wavefunction—describable only with complex numbers: only at the very end does the composite collapse into a real number by squaring. This squaring has an amplifying effect on how probabilities combine in the new physics.

This is an illustration of how increasing the size of a probability amplitude by a modest amount increases the probability exponentially:

$p \longrightarrow p^2 \longrightarrow 1\,P$

$2p \longrightarrow (2p)^2 \longrightarrow 4\,P$

$100p \longrightarrow (100p)^2 \longrightarrow 10{,}000\,P$

$1000p \longrightarrow (1000p)^2 \longrightarrow 1{,}000{,}000\,P$

What if this internal amplification applies to our linear chain of aminoacids and thousands of water molecules? The quantum probability would be a million-fold, while classical concepts only suggest a thousand. As a wage earner, I am all-too-aware of the difference between a thousand dollars and a cool million.

Many of the "not-common-sense" phenomena in the quantum world are a consequence of this internal amplification: we just don't expect probability to behave in this exponential way. (If Lady Luck behaved this way, buying lots and lots of MegaMillions tickets would actually make a great deal of sense. But she doesn't.)

Principle of Least Action

Naturally, a question comes to mind; where does this intangible probability amplitude come from, what is its cause. We mentioned that Planck's constant involved "the action"—and this is where the wavefunction comes from.

Quantum physics considers action of fundamental importance. "The fundamental law of quantum physics states that the probability amplitude of a given path being followed is determined by the action corresponding to that path."[1]

The probability amplitude, measured by complex numbers, is like a tendency to follow a particular history. This tendency is consequence of the action along that particular path, a tendency with a size and a direction "pointing" in an internal dimension, not an external one of time or space.

The connection between the size and direction of this internal tendency and the action along a path of history involves a somewhat sophisticated mathematical construct called a path integral. We need not go into this in any detail, thankfully.

Briefly put then, we can say that the cause of quantum probability form—itself the cause of the probability of what will happen—is the Principle of Least Action. This is the basic law of the universe and is the ultimate cause of what happens in the universe eventually.

To summarize:

In classical science, the action determines what happens

In the new science, the action determines the *probability* of what happens.

It's a subtle, but highly significant difference.

It allows, for instance, the concept of autonomy to appear in science at the level of subatomic particles.

Autonomy

Neither the internal nor the external probabilities are directly observable. We do not observe probabilities, after all, we witness events and interactions.

---

1 A. Zee, Fearful Symmetry: The Search for Beauty in Modern Physics, Macmillan, NY (1986), p. 142.

This is the final step in the new physics. The step from the probability of an event happening to the event actually happening. This is the collapse of the wavefunction into an observable state.

This is a basic question asked by scientists: "What determines what actually happens?"

This has a simple answer in classical science: Natural laws determine what happens.

In the quantum world, however, natural laws determine probability, and only probability; nothing more, nothing less.

Natural law, in the new science, does not determine what happens. Surely, our classically-raised minds complain, there must be something determining what actually happens, some process that can be described by mathematics, no matter how sophisticated that might be.

Quantum physics takes a quite unexpected turn here: it asserts that the connection between probability and actuality cannot be pinned-down in an equation that predicts what happens.

There is a mathematical description of this step in the new physics, of course, but it is another non-classical concept, that of the random choice operator.

This is basically total randomness, which, almost by definition, cannot be described by an equation or a program.

Controversy about the "meaning" of this failure to take responsibility in the quantum view abounds, on a par with the "meaning" of the little arrows pointing internally. They, at least, can be described by definite equations.

We have already seen that the quantum view is different in that it introduces quantum probability forms, QPF, into the mix.

This is another big difference between the two views: Quantum physics has dropped the concept of determinism—there is no well-defined aspect of reality that determines what happens, given the probabilities.

Quantum science avers that there is nothing, other than the probability, that determines what happens externally. In dealing with this we will encounter concepts that, to the classically-trained, are almost as wrong-headed as admitting that matter has an internal extension.

To hearten the reader through this classically-disturbing section, I will mention two points surrounding the controversy about the "collapse of the wavefunction" as this transformation from probability-of-happening to actually-happening is often called:

• Scientists only feel panic about "what it all means" when thinking in regular language. The mathematical formulation is flawless—the problem is translating the perfect-description math into imperfect everyday language. As we shall see, however, the math required to describe the collapse of the wavefunction is less challenging than complex numbers—and I hope these numbers feel as natural as -1 by now. So, while by the end of this section we will be forced, by language, to use such provocative terms as "autonomy" and "free choice," hopefully the mathematical concept of a "random choice operator" will be associated with the words and not some vague philosophical or cultural concepts with all their attendant baggage.

• The failure of predictive ability is really only a problem when dealing with a small number of events. When huge numbers are involved—as they often are in most natural systems we will be looking at—the Law of Large Numbers takes over, so to speak, and ends up turning probability into actual history. This is a well-defined formulation, there are lots of nice, well-defined equations involved. So, for most natural systems where lots of interactions are involved, the external history is determined by definite equations which do predict what happens. When lots of interaction occurs, the quantum view is just like the classical—overall history is determined.

So, if the reader finds the following section difficult to digest, worry not; its concepts will be rarely invoked as we progress up the hierarchy of matter to the new-physics description of atoms and living systems. Though it does have a role to play in the origin process as we shall see later.

COLLAPSE OF THE WAVEFUNCTION

The connection between probability and what actually happens is called the "collapse of the wavefunction." In the slit experiment, we ended with a precise method of calculating the probabilities of the detectors firing in a slit experiment. But the slit experiment does not deal in probabilities; it deals with detectors firing. We take the final step of describing what determines which detector will actually fire.

Here is one of the most greatly unexpected concepts in the quantum view of the world: that there is absolutely nothing whatsoever that determines what happens in this final step.

Shoot a solitary electron through a 2-slit apparatus. We know how to calculate the probability amplitudes involved so can calculate the probability of a detector firing; but for this solitary electron we would like to know which detector will fire.

It turns out that this is an unknowable for it is firmly established in quantum physics that it is impossible to predict which detector will fire. Put another way, experiment insists that there is no natural law that determines which history the electron actually follows. In the new physics the laws all work on the internal level: there is no law governing the external.

QUANTUM RANDOM OPERATOR

Describing such random choice is difficult in mathematics, almost by definition. This inability is concealed somewhat in the official-sounding quantum random operator.

Experiment insists that nothing determines which path the electron will actually follow: which one it will "choose." The behavior of an electron is totally indeterministic—sometimes it will "choose" to go A, sometimes B, and nothing can predict which one. One path is chosen and the other is relegated to the realm of the might-have-been. This is why we can relabel the collapse of the wavefunction as "autonomy of choice" from the particle's point-of-view.

The math of the random operator's somewhat disreputable roots lies in gambling (sorry, gaming) theory—and is simplest to discuss in terms of "equal-probability and proportional representation." It sounds worse than it is. If you have ever tossed a coin for gain, you already know all you need know.

"Each aggregate describing all possible outcomes ... would be called a sample space .... In general, a sample space of a random experiment is a set of elements such that any outcome of the experiment is represented by one, and only one, element of the set."[1]

The operation that picks one of these from a set in game-theory mathematics goes under the name random choice generator. That's basically it—the quantum random choice operator. The math description of the choices made during the wavefunction collapse is just that simple and uninformative.

What the operator-description of the collapse of the wavefunction means is probably the most contentious issue in the philosophy of quantum physics. This debate of meaning, by the way, is almost irrelevant to most working scientists: the math of quantum mechanics has performed flawlessly under the most challenging of experiments. The worst fate that can befall the quantum description is, like Newton's, to be found to be an approximation of a more sophisticated reality. But wrong: never. So, the lack of consensus on meaning is not troubling while churning out the correct answers.

"The orthodox Copenhagen interpretation of quantum theory is silent on the question of the collapse of the wavefunction. The field is therefore wide open .... Any suggestion, no matter how strange, is acceptable provided that it does not produce a theory inconsistent with the predictions of quantum theory known to have been so far upheld by experiment. Our choice is a matter of personal taste."[2]

---

1 Edna E. Kramer, The Nature and Growth of Modern Mathematics, Princeton (1970), p. 259.

2 J. Baggott, The Meaning of Quantum Theory, Oxford, (1992), p. 185.

An extreme example of the random choice operator in action is that involved in the decay of a uranium atom by emitting an alpha particle. The probability that an alpha particle will make it out of the nucleus is extremely low. The nucleus, you see, behaves just like a liquid drop with a surface tension. A little ball of mercury escaping from a broken thermometer used to be a common example of this, the surface tension pulls the liquid metal into almost a perfect sphere. Now the surface tension of the nuclear 'fluid' is trillions of times that of mercury—it is not easy for a tiny drop of fluid (the alpha particle) to overcome this inward pull—there is a potential energy "wall" bounding the nucleus. The alpha particle collides with this wall trillions of times a second and usually bounces right back.

It takes, on average, hundreds of billions of years for the alpha to teleport across the wall and escape. Each bounce at the wall—each attempt to escape—is an event. Each event involves the random choice operator. In the equal-probability, proportional representation description of this operator, it is picking with equal probability an item from a huge set of probable histories that looks something like this:

1 : 1,000,000,000,000,000,000,000,000,000,000,000

While this seems odd and somewhat simplistic, the math does not get much more sophisticated than this—the key characteristic of absolute randomness, after all, is that it cannot be described by an equation. The math is telling us there is no describable process happening there—randomness is the absence of something defined.

Unlike everything else in the new physics, the random choice operator is difficult to describe in mathematics. It is impossible, for instance, to formulate an equation that has a random solution. This is why randomness is difficult to model on computers even though it would seem, at first thought, to be simple. The problem is that the generation of random numbers is not handled at all well by computers—programming a computer to come up with a truly random sequence of numbers is impossible. The best consequence of classical true-or-false logic in which "random" can only be approximated—pseudo-random numbers, as these best approximations are called.

APPARENT DETERMINISM

Indeterminism has given many a theoretician difficulties, one of which is, where did determinism go? This challenge to the quantum perspective is the reverse of the old: Why do so many systems in nature seem to have no freedom? Why are material systems so predictable and apparently ruled by law?

Even on a fundamental level, apparently deterministic laws such as "light travels in a straight line" and "light travels at the speed of light" are now understood as an artifact: photons actually have a probability amplitude to travel faster than the "official" speed limit and to veer off that way rather than toe the line, but these tendencies are canceled out by the tendency to go slower and veer this way—the only tendency that does not cancel out is the tendency to travel in a straight line at the speed of light. "The amplitudes for these possibilities are very small compared to the contribution from speed c; in fact, they cancel out when light travels over long distances. However, when the distances are short ... these other possibilities become vitally important and must be considered."[1]

If, at a fundamental level, nature is indeterministic, where does the apparent determinism come from? Colloquially put, how can "diamonds be forever" if the electrons and quarks they are made of are free to do their own thing?

"Einstein ... spent much of the rest of his life looking for the deterministic clockwork that he thought must lie beneath the apparently haphazard world of quantum physics. The clockwork has not been found. It seems that God does play dice."[2] Our everyday experience, however, is that "God doesn't play dice"—things in nature seem very ordered and predictable.

---

1 Richard P. Feynman, QED: The Strange Theory of Light and Matter, Princeton University Press (1985), p. 89.
2 P. Davies & J. Gribbin, The Matter Myth, Simon & Schuster, NY 1992.

This is probably just as well for the development of science, for the fact is that most simple systems, the ones studied by physicists and chemists, are not at all stochastic; they are quite predictable and amenable to being described by simple laws. This convenient state of affairs arises for reasons that can be roughly classified into "multiple choice" and "no choice."

MULTIPLE CHOICE

Systems made of electrons can seem to be deterministic and predictable because the number of electrons involved is so large and they are all in similar states. The Law of Large Numbers (LoLN) applies and what happens externally will absolutely reflect the probabilities.

A simple example of this as it applies to autonomous electrons is the deterministic behavior of electric current that "obeys" the deterministic law:

voltage = current x resistance

This simple relationship underlies much of our civilization. The reason for such absolute predictability is that a current of even a few micro-amps involves quintillions of electrons in very similar states all moving in a QPF called a conduction band. Even though the path a single electron takes through a metal cannot be predicted, the LoLN ensures that the behavior of the whole swarm accurately expresses the probabilities.

The indeterminacy of the single electron is lost in the predictability of statistics. (This, of course, applies to the most sophisticated of systems—the success of insurance and mass-marketing institutions is an indication that even we humans are ruled by probability and are more statistically predictable than we like to admit.)

NO CHOICE

Even on the individual level, such as an electron participating in an atom, the behavior seems to be deterministic. This occurs because, while the electron is free, it has a strictly limited set of probable paths to choose from, all of which happen to lie within the atom. The set of states is bounded.

In the simplest situation, if all the possibilities have a zero probability except for one that has 100% probability, then autonomy has no choice: it has to pick the path with 100% probability. This, for instance, is the case for high-energy photons, electrons and the like: they have a 100% probability of plowing ahead in a straight line.

The way to make an electron do what you want, therefore, is not to look for some external law which will force it to your will, but to arrange things so that the probability of it doing what you want is 100% and the probability of it doing anything else is 0%.

An example of this is the ubiquitous TV tube. Here high-energy electrons are manipulated such that the probability of them illuminating the correct pixel is 100% for all intents and purposes. Of course, a lot of electrons are used, so the above multiple choice situation is exploited by the TV designer.

Electrons in stable structures, such as a helium atom, retain their self-determination, but the choice is limited to being close to the proton. All the possibilities that lead the electron away from the proton have zero probability, and the electron never goes that way and helium is stable.

From this lack of choice comes the stability and structural consistency of the world around us. Electrons do not get to exercise their autonomy in a chaotic fashion, not because they have lost their autonomy, but rather because the probabilities are constrained by the electron's environment. In this way, the autonomy of particles is constrained and complex stable structures can be constructed out of them.

The great chemical differences—and the structure of elderly stars—turn out to be an example of freedom-but-no-choice, for the probability of two electrons being in exactly the same state is zero; it is impossible and never happens.

The reason why the 96 electrons in a uranium atom don't bump into each other is simple: they have a zero final probability amplitude to do such a thing and thus, with zero probability, don't do it. Electrons don't zip around—and thus perhaps bump into another—they jump around without traversing what lies in-between and, the key point, they have a zero probability of landing anywhere close to another electron—the probability amplitudes just add up this way.

It is these restrictions that the electrons impose upon one another that make the solution of the wavefunction equations so difficult in atoms other than hydrogen (where there are no other electrons to take into account) and currently impossible for atoms such as uranium.

TIME AND PROBABILITY

There is a useful relationship between the probability of something happening and the time involved in it actually happening—simply put, probable things happen sooner than improbable ones. In a few situations, the theoretical structure developed in a science is mature enough to directly calculate the probable future of a system. Such is the case, for instance, with the electron.

In other situations, the probable future can only be determined by observation—but probability, of course, is not directly accessible; it has to be measured. A useful measure of probability takes advantage of the LoLN and the larger the number of attempts, the closer the results will reflect the probabilities. And if the attempts are spread out in time, the more time there is involved, the more attempts will be made. The actual history over sufficient time will accurately reflect the probabilities.

So, what is sufficient time? In radioactive studies, this period is called the half-life, the time it takes for 50% of a large number of radioactive atoms to decay. This gives a very useful measure of probability in terms of time—the period of time in which the system has a 50% chance of changing.

A neutron, for instance, is a triplet of quarks in a certain state and one of them can change. When it does, the neutron becomes a proton. A collection of 4 quintillion neutrons has only 2 quintillion left after about 12 minutes, 1 quintillion after 24 minutes, etc.—its half-life for the change, rounded up, is 12 minutes. The probability of a neutron changing into a proton over this period of is 50%.

As the word "lifetime" is unsuitable as a general measure of the probability of change in state—it connotes breakdown—we will use the more general term "transition time" and this is related to the probability that the transition will occur. As the transition time increases, the probability decreases, so the probability is inversely proportional to the transition time. This allows us to use the transition time as a measure of probability.

ESTIMATING PROBABILITY

There are two situations in which the actual history fully expresses the probable history: a large number of systems in the same state or a single system iterates the same state. This is the Law of Large Numbers at work.

When there is just one system being observed, the LoLN can give no help, but an estimate of abstract probability by concrete measurement is still possible. In this case, there is only one measurement, the time it actually takes for the change to occur. For instance, although the half-life of a neutron is about 12 minutes, there is a probability that it could go for a whole year without decaying—and at the end of the year there would still be a 50% chance of the neutron falling apart in the next 12 minutes.

Such longevity, however, is extremely unlikely. Statistical theory asserts that it is highly probable that the single measurement will fall within a certain distance of the "mean lifetime" which can be considered the probable time period in which the system will change. Each of the italicized phrases has a defined meaning in probability theory—for instance, highly probable is usually set at 95, 99 or 99.9 percent of the time, depending on the situation.

The two measures have a simple connection derived by the math of probability statistics:

mean lifetime = transition time x 1.44

This relationship between the probability of the change in state (measured by the transition time) and the observed time of the change will prove useful in the discussion of "origins" when only one, or just a few, systems are involved. If only one measurement of an event is made, a good estimate—to the accuracy given by probability theory—of the probability of the change in state is

transition time = observed time/1.44

A provocative use of this relationship, which we will mention here and return to later, is that current estimates of the time taken to go from an abiotic earth to the emergence of the triplet code and the prokaryote-level of life forms was about 100 million years. A good measure, then, of the probability of this occurring is that the transition time is about 70 million years.

This contrasts dramatically with calculations based on classical concepts of random aggregation which give the probabilities of even moderately-complicated proteins emerging—let alone the sophisticated constructs necessary for the functioning of the triplet code—in times vastly in excess of the 15-billion-year age of the universe. Something is clearly missing in these calculations based on classical random chance-and-accident principles.

ARROW OF TIME

The movement through time has always been a bit of a puzzle in science. Our common sense division of time into past, present and future has no basis in classical science. The laws are reversible; they apply equally well to time "running backwards"—moving in either direction along the world-line is possible according to the classical perception. So where does the one-way nature of time come from?

"The laws of science do not distinguish between the forward time and backwards direction of time. However, there are at least three arrows of time that do distinguish the past from the future. They are the thermodynamic arrow, the direction of time in which disorder increases; the physiological arrow, the direction of time in which we remember the past and not the future; and the cosmological arrow, the direction of time in which the universe expands rather than contacts."[1]

As human beings, we are mortally aware of the difference between the past and the future:

The Moving Finger writes; and having writ, / Moves on: nor all thy Piety nor Wit Shall lure it back to cancel half a line, / Nor all thy tears wash out a Word of it. [2]

The problem with classical physics was that it could not establish this arrow of time—the equations of motion work just as well for time running backwards.

The past state and the future state look the same to the classical equations which determine what happens to the external extension of a system. You can run this "movie" of reality backwards and it still makes sense.

The laws of nature are all, fundamentally, time-reversible. This symmetry is easily illustrated by replacing the present with a mirror—reflecting either the past or future in the mirror does not change the look of things—the world-line is symmetric to time reversal.

Another way of looking at this is that if time were to run backwards the same laws would be apply. This is true for all natural laws: both the externally-acting laws of classical science and internally-acting laws of quantum physics.

The laws of the new physics are time-reversible—they apply equally well to the internal extension whether time is running forward or backwards—it's the same action equation after all.

But once we bite the bullet and accept the random choice operator, we have a loss of time reversibility. The random choice operator makes no sense running in reverse: how can you add an actual history back into a set of probable histories? It makes no sense in reverse. The description is no longer time-reversible—you can't run the random operator in reverse—unlike the other operators which do make sense in reverse.

---

1 S. W. Hawking, A Brief History of Time, Bantam, NY (1988), p. 152.

2 Edward Fitzgerald, The Rubáaiyát of Omar Khayyám (1859).

The full quantum picture of the past, present and future is not symmetrical—the reflections in the mirror no longer look the same.

The past-to-present is external and singular—the path the system just chose, while the future is internal and plural—the paths it has a probability of following. The present is the time when the random choice operator does its work. This does not look the same when a mirror is inserted.

This is the fundamental difference between the past and future implicit in the new physics. This aspect of quantum physics actually appeals to our basic sense of the difference between past and future—the moving finger has a single past to repent of what it has writ, but there are multiple possibilities for what it might write in the future.

This arrow of time appears naturally in the quantum view for, while both classical and quantum views have deterministic natural laws, the new physics has these laws acting on the internal extension, which can be multiple, and not the external extension, which is singular.

The only reason that this in-built arrow of time has not been fully recognized is that thinkers have usually taken the route of thinking of the internal extension as not really "real"—as a mathematical tool, but nothing more. If the internal extension is just a mathematical fiction then, of course, the in-built asymmetry in reality of past, present and future is no longer apparent. We have avoided a lot of difficulties and taken the path of least resistance in accepting that, if science insists on including an internal extension in its description of the bricks and mortar of reality, then we might as well take things at face value and accept that the internal is just as real as the external.

Simple Schrödinger

Now, when someone mentions Schrödinger's Equation at a dinner party (I wish), you can say:

"Oh yes, pixels and probability. So simple, really, that the negative steepness to the gradient has to fit into itself. I'm just fascinated by those little 'h' squares of chess jumps to existence. And I just 'sigh' over the wavefunction, don't you. So enigmatic. All those little cupid arrows pointing in not-spacetime. Sounds like my minister giving sermons about sin as "hooks" that stick out in spirit world and catch on more sin. Wonder if they are pointing in the same abstract space. And that final conjugation of the wavefunction into rough and ready probability is so romantic, somehow. Where would we be without that cute little 1s orbital, I ask you? Nowhere in a body, that's for sure!"

$$-\frac{d^2\psi}{dx^2} = \frac{2m}{\hbar^2}(E - V(x))\psi$$

That should lead to calls for "More Wine!" around the table.

By the way, just in case asked, the rest of the equation is boring stuff from high school: like mass and distance from the center, and electromagnetic energy penduluming from kinetic to potential and back.

See, once you know how to translate appropriately, math is not so hieroglyphic after all.

We looked at this equation in detail for one simple reason; it is basically as far as the quantum revolution has reached in terms of a rock solid mathematical equation. And it can barely be solved—not for quantum reasons, but because a precise classical-aspect calculation of the penduluming from kinetic to potential energy in multi-electron atoms is beyond current techniques.

You may breath a psi of relief; there are no more quantum equation-hieroglyphs to deconstruct.

Absolute Power

To my way of thinking, the most remarkable concept established by the new science is the awesome power-over-matter ascribed to the wavefunction, the supremacy of quantum probability over common sense notions.

The concept of probability having power over matter is gobbledygook in classical science where probability it is considered a lowly result of a lack of knowledge, if anything.

This is not so in the new physics where quantum probability plays a central role in the conceptual framework of the new physics. Thus, the quantum probability of two electrons being in the same state is zero. Now this is not the almost zero of everyday life, or even the calculus; but exactly zero. The power of probability is so absolute that all scientists are quite confident that not a single electron in the entire span of the universe, in the entire past and future of the universe, has ever been in the same state as another electron. Never has; never will; Verboten.

The power of this quantum impossibility—which in math-speak is almost as simple to calculate as one minus one equals zero—is so profound that it can hold up an entire star without any assistance. So, in a billion years or so, when our sun runs out of hydrogen to burn, it will collapse under its own weight until it has shrunk a million-fold. At this point, however, the electrons will be on the verge of being forced on top of each other, to share the same state. As this has a zero probability of happening, the sun will abruptly stop shrinking and become a stable white dwarf. All that is holding it up against the lash of a billion gravities is the power of quantum probability. An exhibition of Power that even Superman might marvel at.

In the following discussion is important to keep in mind that quantum probability is fundamentally not at all the same as classical probability—the coin toss-MegaMillions variety we are more familiar with. Rather, it involves sophisticated concepts such as complex numbers, probability amplitudes, orbitals and wavefunctions. As we are dealing with broad strokes, however, we need not delve into the details.

There are two major differences between the two concepts that have to be kept in mind during the following discussion (as the classical concepts so laboriously learned will reassert themselves on the unwary):

Quantum probability is causal and measured with complex numbers • Classical probability is resultant and measured with real numbers.

Quantum probability forms are discrete and relatively few in number • Classical forms are continuous and multitudinous in number.

QUANTUM INTERACTION

Classical and quantum theory both basically agree on the external aspect of what happens when systems interact: they trade, exchange and share bits of themselves with other systems.

In the last section on the structure of systems, we focused on the subsystems that were tightly held. Now we will deal with those with a tendency to stray.

"In the physical realm, operations arising from the interplay of four forces are transmitted by messenger particles.... In the biological realm, operations... are transmitted by messenger molecules.... This correspondence reveals a fundamental program of nature...."[1]

There are three things that have to be taken into account in this trading or coupling with subsystems. We have established that a system has:

1. Subsystems in quantum probability forms with intense gradients that have no tendency to stray. If things "fall into" probability then these QPF are deep and with steep walls. In the atom, examples of these are the protons, neutrons and inner, contented electrons of the massive atoms.

2. Subsystems in quantum probability forms (QPF) that have a tendency to leave the system. In an atom, these are the outer, valence electrons.

3. Unoccupied QPF that have a tendency to offer occupancy to passing stray bits that have taken off from other systems.

Only the last two categories are significant in interaction.

Coupling subsystems are those occupying a QPF with a probability of taking off from the system. This is a subset of all the subsystem QPF.

In an atom, for instance, this tendency to lose electrons is called the electro-positive valence. The unpaired, single electron in the atom of lithium is a good example. This lone electron is easily lost when interacting with other atoms. On the other hand, the lithium atom has a very low tendency to take in extra electrons.

The empty QPF with a significant probability of taking up a passing subsystem into the structure of the system is a set of taking-in tendencies, the negative coupling capacity of the system.

The overall capacity for interaction by both giving and receiving subsystems is the combination of these two, a subset of the system's QPF.

This coupling capacity determines what we can call the sophistication of a system, the ways in which it can interact.

CORRELATION

This tendency to interact has consequences if the system is not an isolated one; if there are other systems around to actually interact with. The environment in which the system finds itself, its milieu, also has a tendency to interact. A system and

---

1 Edward Rubenstein, "Stages of evolution and their messengers," Scientific American (June 1989), p. 132.

its milieu, considered as a single system, have the same basic structure of occupied and unoccupied QPF, some of which are open to trading and exchange.

Both system and its milieu have their particular tendencies to interact. From this internal QPF comes the probability of subsystems shuttling about. This can happen in two ways:

The system has a tendency to give out a subsystem that is matched with a tendency of the environment to take that subsystem.

The system has a tendency to give matched with a tendency of the environment to receive.

The give-out QPF of one system merges with the take-in QPF the other creating a path of least action for the subsystem to slip along. Both directions taken together give the overall tendency to interact, the QPF for the interaction.

This is the interaction quantum probability field, a QPF.

INTENSITY OF INTERACTION

From this internal interaction QPF comes the external probability of the subsystem actually skipping on over rather than doing any of the other things it could probably do. The number of coupling subsystems making it over per unit time, the intensity of the interaction, will be proportional to this probability. The more probable the trade, the more intense is the trading.

While the random choice operator will have its say here, the law of large numbers is usually there to cancel its influence over the long run.

In the long run, the external 'form' or intensity of the interaction will reflect the internal form of the QPF.

It is not as complicated as it all looks as an example will show. Take two atoms, one of lithium, one of fluorine. In our step diagram, the QPF orbitals, some occupied, are:

Their coupling capacities are quite different:

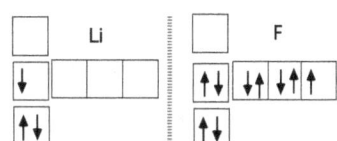

Lithium

a high probability of the singleton electron leaving

a zero probability of taking in another electron

Fluorine

a zero probability of any electron leaving

a high probability of taking in an electron

The correlation of fluorine giving out and lithium taking in is 0. This route makes no contribution to the interaction.

The correlation of lithium giving out and fluorine taking in gets the extra quantum boost when adding probabilities we encountered before.

Just how far the electron jumps as the two atoms approach is not recorded, but it's fast. The electron leaps from the lithium to the fluorine and everyone is happy; all the electrons are now in contented pairs. The result is two ions clinging to each other, the so-called ionic bond in chemistry.

Carbon and hydrogen are not as extreme as these two; they have about equal tendencies to give out and take in. They end up sharing contented electrons in pair-bonds. These bonds are the sticks holding the balls together in classical models of molecules.

FUNDAMENTAL INTERACTIONS

That interaction involves coupling was not that obvious at the lowest levels of the material world. Rather, classical physics described interaction in terms of "forces" acting on material systems, some by direct contact like balls colliding, some at a distance like gravity.

In 19th-century physics, systems were thought to interact through the mediation of abstract, intangible fields of force, such as gravity or the electromagnetic field. In the 20th-century view, these fields were understood in terms of probability of coupling with subsystems. The four fundamental interactions of physics involve coupling with particles. The field equations of modern physics describe the probability that the coupling with subsystems will occur. First, we will take a look at the phenomenon of coupling and then at why coupling creates what classical science calls forces.

Experiments reveal that the universe contains just two types of fundamental particles—called the fermions and the bosons—which have been likened to the "bricks and mortar" out of which everything is constructed. While both virtual fermions and virtual bosons are to be found in the description of the quantum foam of the last chapter, it is the presence of the bosons that has the greatest implication for interaction.

The bricks of the material world are the fermions, the "bits of matter" such as the electrons and quarks. These fermions have the rather odd characteristic property—called "spin half-integral"—of needing a rotation of 720° to return to the same orientation. This otherwise rather mysterious property has been interpreted as support for Superstring Theory in that it is topologically equivalent to behavior on a Moebius strip [1]—a twisted surface on which it takes two circuits to return to the original orientation.

The mortar is the interaction between these fermions which involves an exchange of bosons—particles with the more familiar property, called "spin integral," of needing just a 360° rotation to return to their original orientation. It is the exchange, or "coupling," with bosons that unites the fermions together into composite structures.

Basic Interactions

Such coupling with exchange particles lies at the heart of the four basic interactions (or classical forces) known to physics: gravity, electromagnetism, the "strong" and the "weak" nuclear interactions. The best-characterized of these is the electromagnetic interaction where the bosons are the photons (particles of light) and the fermions are the electrons and quarks, both of which have "electric charge."

Electromagnetic Interaction

In classical physics, electric charge was something a particle had and the electromagnetic interaction was described as an action at a distance through electromagnetic fields.

One of the many reversals that occurred in the development of quantum physics was the realization that electric charge is not something a particle has but rather something a particle does—charge is simply the tendency of a particle to absorb or emit photons. To say that a particle has an electric charge means exactly the same as saying that it has a distinct tendency to absorb or emit photons—it "couples" to photons. This coupling is not an electromagnetic interaction—the photon itself has no charge—and exactly what is going on as an electron and photon merge or separate is a mystery since the structures of both of them are unknown.

Particles with "charge" emit and absorb "virtual" photons. Virtual photons do not suffer the time restrictions on virtual electrons since they do not experience, so to speak, the passage of time. Einstein's Special Relativity revealed that the faster you travel the slower time passes until it stops altogether at the speed of light. While from our reference frame, it takes a photon of light 20 billion years or so to cross the visible universe, in the photon's reference frame it takes no time at all.

So, during the brief existence of a virtual photon—which, of course, travels at the speed of light—it can actually travel an infinite distance. On its travels, the virtual photon can be absorbed by other electrons or quarks—coupling the particle that emitted it with the particle that absorbed it—giving the electromagnetic "force" an effectively infinite range. While it is a geometrical requirement that the "density" of these virtual particles falls off with distance, it is never, no matter how far the

---

[1] E. E. Kramer, The Nature and Growth of Modern Mathematics, Princeton Paperbacks (1970), p. 610.

distance, exactly zero. Interaction is usually described in terms of fields. The intensity of the electromagnetic field depends on the probability of encountering virtual photons at a location.

PLUS AND MINUS CHARGE

In non-quantum terms, the electron has a "negative" charge and the proton a "positive" charge, in the convention established by Benjamin Franklin. The difference between these charges does not reside in the capacity for the emission and absorption of photons—for either charge, the tendency to emit a photon is always exactly equal to the tendency to absorb one. In fact, when two particles interact electromagnetically, the uncertainty inherent in subatomic systems makes it impossible to know which system did the emitting and which did the absorbing; all that can be said is that a photon was exchanged.

The tendency of a plus or minus charged particle to couple with a photon is called its "coupling constant" and it has a value of about 1/137 for the charge on the electron. The difference between plus and minus charge is related to a type of polarization. While the exchange of virtual photons does not transfer energy between particles, it does transfer momentum. For particles with the same "charge," such as two electrons, the polarization of this transfer results in the electrons moving away from each other, just as it does when two positive charges couple. For particles with opposite charge, such as a proton and electron, this transfer moves them towards each other. We will return to this point when we, eventually, get to the discussion of the consequences of coupling with subsystems.

QED

The electron and proton in an atom interact by coupling with a prodigious number of photons—classically speaking, there is a powerful electromagnetic force between them.

The theory that describes this exchange is Quantum Electro Dynamics (QED). For all the staggering complexity of the actual calculations, the underlying structure of the QED equations is simply an iteration of two tendencies:

1) the tendency of a charged particle to absorb or to emit a photon.

2) the tendency of a photon to move from one place to another.

In modern physics, these tendencies—or, more technically, probability amplitudes or, more simply, internal probabilities—are at the root of all phenomena involving light and electrons—which embraces just about everything except gravity and the structure of the nucleus.

The electromagnetic force as a classical "force at a distance" working through an abstract force field is not part of modern science; instead, it is now understood as a substantial exchange of particles.

It is true that we usually do not think of an electron as having photons inside itself—after all, photons are huge compared to the size of the electron (or atom, for that matter). Yet, photons do definitely emerge from electrons as well as disappear into them. The vacuum foam of virtual particles obviously has to be included in the list of what a system is composed of. So, ignoring the outrage engendered by the thought that systems can contain things bigger than themselves, we shall include the photon—as well as all the virtual particles we will shortly encounter—in the substructure of the electron (and other particles).

The three other "fundamental" forces are described in the same way as the electromagnetic.

THE WEAK NUCLEAR FORCE

Particles that "feel" this force (they are said to have a "weak charge") couple with particles called the W and Z intermediate vector bosons. These were predicted by theory, then detected and are now 'factory-produced' by a team of European scientists. They are now being produced in quantity in at least two high-energy facilities.[1] These weak bosons, like photons,

---

1 R. Lewis, "The Three Families of Matter," The World & I, April 1990, pp,. 300308.

are emitted in a virtual form, travel to other particles and are absorbed—the effects of this exchange being what we call the weak nuclear force. Unlike the massless photon, however, the bosons are massive and, traveling way below the speed of light, are all-too-mortal, falling apart in time measured in trillionths of a second. This ponderous mortality severely limits the scope and effect of the weak force and also accounts for its name.

The weak interaction plays a role in changes within the atomic nucleus. It has little to do with everyday life except for the essential role it plays in moderating the first step in the fusion of hydrogen to helium that powers the sun and, ultimately, all life on this planet.

THE STRONG NUCLEAR FORCE

Particles that "feel" this force couple with gluons. While gluons and quarks cannot be isolated—they are "confined"—and can only be detected indirectly, both types of particles are considered firmly established in quantum physics. The confinement of the gluons limits the effect of the strong interaction to within the nucleus, but there it is hundreds of times stronger than the electromagnetic interaction.

The strong interaction is described by "Quantum Chromo Dynamics" which mirrors the equations of QED: the quarks making up the protons and neutrons in the nucleus couple to "gluons" which bind the quarks together. Analogously to electromagnetism, quarks are said to have a "color" charge, though now there are three types of polarization whimsically called red, blue and green.

Gluons, unlike the electrically-uncharged photons of the electromagnetic force, couple to themselves; they have "color charge." This is one of the reasons why QPF is much more complicated than QED—another being that there are three charges and the coupling constant is close to unity.

During the period when physicists were taking a good look at the strong nuclear force, high-energy colliders produced hundreds of particles. They were real but unstable. Many of these had lifetimes so short that they could not be easily detected directly and were instead observed to be "a resonance" spike in a graph of the results.

Note, that while these lifetimes are brief indeed—on the order of 10^{-24} of a second before disintegrating—the atomic nucleus is so small that these ephemeral particles, moving at a fraction of the speed of light, live quite long enough to make a few circuits of the nucleus. Thus, as virtual particles, they all had to be taken into account if the strong force that holds the nucleus together is to be properly understood. The coupling substructure of the proton and neutron thus include a large number of these ephemeral resonance-particles and thus have to be included in the list of subsystems available for coupling with.

Thus, while the positive protons in a large nucleus are copiously exchanging photons and experiencing an intense repulsion as a consequence of the momentum exchange they, along with the neutrons, are also exchanging copious numbers of all sorts of virtual particles—a resonance which exerts an exactly opposite effect and pulls the nucleus together. The "strong force" between protons and neutrons is based on the more fundamental strong force between the quarks—just as chemical bonds are based on the more fundamental electromagnetic force between electrons and protons.

Our earlier analogy of Yankee Stadium is no longer large enough. An excellent overview of the inner secrets of the quarks puts the quark size scale in perspective:

In the magnified analogy ... with a human reaching the stars, atoms the size of the Earth, and [the protons and neutrons in the nucleus] fitting inside a playing field ... we can say that a bare quark must be smaller than 10 centimeters, or about two-and-a-half inches, across."[1]

Electrons are about the same size. So, in a hydrogen atom, we have one electron and three quarks about the size of baseballs in a volume the size of the Earth. Clearly, an atom is a lot of orbital, a little bit of stuff doing the 'filling in.'

---

1 Timothy Paul Smith, Hidden Worlds, 2003, p. 50.

GRAVITY

Even though a quantum theory of gravity is not established, the gravitational interaction is also thought to involve coupling with hypothetical particles called gravitons. This concept, however, introduces a schizophrenia into modern physics since Einstein established the phenomenon of gravity as a curving of space-time, a bending that is mild in the vicinity of a star such as our sun but can be intense enough to "pinch off" a piece of space-time—as happens in the formation of a black hole.[1]

There is a growing consensus that at very high energy—such as abounded in the moments after the Big Bang—the differences between the particles—electrons, quarks, photons, weak bosons, etc.—disappear. These are the Grand Unification Theories and Theories of Everything that are at the cutting edge of modern physics. One of the more successful of these, the Superstring Theory, suggests that all particles, both fermions and bosons, are the result of an extremely intense curvature of extra exotic dimensions, and that regular gravity and gravitons are just a pale echo of this more fundamental level of reality. It is this sort of convergence of ideas that gives many theoreticians hope that quantum mechanics and gravity can be formulated in a consistent way.

INERTIA

One of the enigmas associated with gravity is the link between inertial mass and gravitational mass. Mass is the measure of a system's capacity to gravitationally attract other systems; inertia is the measure of a system's reluctance to change velocity. There is (currently) no compelling reason why these should be the same, yet they are, to the accuracy of the best measurements.

Einstein avoided the problem for linear motion by removing the concept of absolute motion in the Special Theory and by postulating the equivalence of mass and inertia as a foundation for the General Theory.

Unlike the validity of absolute linear motion, the validity of absolute rotational motion is still under debate. There is a protracted and confused debate, which continues to this day, as to whether Einstein's General Theory of Relativity does or does not incorporate Mach's Principle: That there is no such thing as absolute <u>rotational</u> motion, only <u>relative</u> rotation. (Similar to Einstein's position on linear motion.) The implication of this, however, is that rotational inertia ("centrifugal force") is caused by the presence of the rest of the universe. Would the earth bulge at the equator if it rotated alone in the universe? No one really knows, but as linear inertia involves the Higgs—see below—we can guess that the answer will also involve them as well.

For all these questions surrounding gravity, both mass and inertia are an expression of graviton coupling. The most useful scientific measure of graviton coupling is actually not mass. While classical and quantum physics have quite different perspectives on what the mass of a system is and how it changes, both agree in placing momentum in a central role. Momentum combines both the gravitational and the inertial into a quantity that is neither created nor destroyed: rather, it is a measure of graviton coupling that is rigorously conserved.

Momentum is the product of mass times velocity. Mass is the measure of graviton coupling. Velocity is change of position in the inertial frame with time. This movement through the combined gravitational field of all the matter is a measure of the inertial gravitational coupling with gravitons.

Momentum, the product of mass times velocity, is thus a conserved measure of the gravitational and inertial interaction of systems with gravitons.

---

[1] Stephen W. Hawking, A Brief History of Time: From the Big Bang to Black Holes, Bantam Books, Toronto (1988), p. 81.

HIGGS

In order to bring us up to the cutting edge of modern physics we will briefly mention the Higgs mechanism. In all of the interactions discussed so far, theory implies that the properties of the couplers should be very similar—clearly wrong as, while the photon and gluon are massless, the weak bosons are not. A mechanism has to be introduced to explain this symmetry breaking, and this mechanism involves particles coupling to the "vacuum" with Higgs bosons, massive particles (i.e., of very short range) that the super collider in Texas was supposed to look for.

It is this coupling to the vacuum that is thought to give each particle—bosons and fermions—its characteristic rest mass which is also somehow involved with gravity and gravitons. One of the key differences involves the quantum concept of spin: fermions have half-integer spins, bosons such as the photon, gluon and W have unit spins; gravitons are predicted to have a spin of two while the Higgs is expected to have spin zero. All this will hopefully become clear when a quantum theory of gravity becomes established and science moves down the hierarchy of matter.

We have included the subsystems in the vacuum foam in the substructure of particles such as the electron and quark. One puzzle is: if all particles have the same vacuum foam around them, why do particles have different abilities to couple with them? The vacuum foam of the quarks contains virtual photons, W bosons, gravitons, gluons and it can couple with all four of them. Presumably, the same stuff envelops the electron, but for some reason, it fails to couple with gluons; it does not feel the strong force. What happened to the virtual gluons in the electron's substructure? The neutrino, embedded in the same vacuum foam, does not utilize the photons we would expect to be there—the neutrino does not feel the electric or the strong force. Where did the virtual photons and gluons go?

See the Appendix and Volume Two for some speculations on the Higgs.

BITS OF SELF

In all of the fundamental interactions just described, we see that the coupling subsystems are drawn from the substructure of the interacting systems—the quantum foam structure of the vacuum in the most basic cases.

Interactions involve them coupling with subsystems. Things interact by exchanging bits of themselves in quantum science.

Just as the "fundamental" particles interact by coupling with the subsystems from their substructure, systems at every level do exactly the same thing: they exchange bits of themselves and thus interact with each other.

Systems can interact by coupling (both sharing and exchanging) with any of its subsystems, though they do not necessarily have a significant tendency to couple with all of them. Here, "significant" implies that the tendency has to be taken into account if the behavior of the system is to be understood. These active couplers are a subset of the external hierarchy of stuff filling in the QPF.

The coupling substructure does vary somewhat with situation—for example, water at $-200°C$ has a different set of tendencies to couple with its subsystems than when it's at $+200°C$. The following general discussion assumes everyday standard-temperature-and-pressure situations.

The list of active couplers is a qualitative description of the external aspect to interaction, delimiting the types of interaction a system can get involved in.

This is what systems do. In a nutshell, they interact by exchanging and sharing bits of themselves with others via QPF. What classical science describes as force is a consequence of the quantum probability of the exchange happening in the new physics.

The Path not Taken

There are consequences to the inherent contingency in open histories. Sometimes, taking one path and not the other can have historical consequences. A potentially good example is our current understanding of why all life uses only the left-form of amino-acids and right-form of nucleotides.

As far as we know, however, there is no reason to think that right amino acids or left nucleotides would not be just as good at working together. One explanation is that there is no good reason why our L-R set-up emerged—it was a contingent step along the way and the random choice operator picked one path from the possibilities. Based on this one event, a whole tree of possibilities opens up that ended up with us.

The other combinations never made it to this step, if they did, we out-competed and extinguished them for they have left no trace. We can diagram this with a simple wavefunction: the random choice operator "picks" the path at each node ending with the left-right connection event. They are the paths not taken in Earth's history. Once this L–R situation became established, it is theorized, it preempted all the resources and prevented any attempt to establish any other chiral combination.

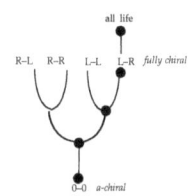

If we could do the calculations (as far as current knowledge seems to predict) there is no reason why a probability that life would develop on the right amino acids and left nucleic bases is also there. a probability, however, is not an actuality. It seems the explanation is contingent; the left-right system appeared first and preempted the stage leaving no probability that another system could develop—the 'contingent evolution' promoted by evolutionist S. J. Gould.

Or, then again, perhaps it will turn out—when we know and can solve the equations that describe all the internal systems and action equations—that a system such as a primate ape is almost certain to emerge over a period of 20 billion years after a Big Bang.

HIERARCHICAL COUPLERS

A hierarchy of interactions is quite simple to list: each level inherits what came before and adds the capacity to couple with its primary subsystems: at each level, the level below contributes an emergent interaction to be added to the inherited ones. Not all of these capacities, of course, are expressed at each level

SYSTEMS	COUPLERS	EXAMPLE
electrons	gravitons, photons	charge
quarks	" & gluons	color charge
nucleons	" & quarks as pions	nuclear force
nuclei	gravitons, photons	weight
atoms	" & electrons	valence
molecules	" & atoms	H bond
macromolecules	" & molecules	water structuring
organelles	" & macromolecules	protein action
cells	" & organelles	fertilization
organs	" & cells	immunity
etc.		

We have now discussed the second concept that the quantum and classical views agree upon—systems interact by coupling with subsystems. Systems are composed of interacting subsystems coupling with sub-subsystems. For example, a water molecule is composed of interacting H and O atoms which couple with electrons and photons.

Primary and secondary

A distinction that will be significant when we get to discussing origins is that of primary and secondary interactions. Put simply, while a system can couple with its primary subsystems, an isolated primary subsystem can only couple with secondary subsystems. The origin of systems deals with the fact that history is filled with examples of times where a system is absent followed by times where it is present. A simple example is the moment thousands of atom-less years after the Big Bang that saw the appearance of atoms. This marked the appearance of a new interaction on the cosmic stage. An atom can couple with electrons, a primary subsystem. Electrons, however, do not couple with electrons, they do not have atomic valence. Only the combination as an atom has valence.

Valence emerges, so to speak, with the formation of the first atom. Valence, involving the primary subsystem is a primary interaction. (Note, that in no way is this to be interpreted as most significant.) Atoms also interact by coupling with photons and gravitons, but these are secondary in that they are inherited from the particles.

We will call the primary interaction the emergent, and the secondary ones the inherited capacities for coupling.

The atom, for example, inherits the electromagnetic interaction from its constituent electrons and protons. On the other hand, the atom does not inherit the color charge of its constituent quarks, so chemists do fine without including color charge in their atomic and molecular action equations (such as they have).

As systems congregate together as a supersystem, the supersystem can start to couple with those very systems doing the congregating. We will return to this point when we get to the Origin of a system after we have figured out what makes the systems do the congregating in the first place.

The atom can couple with electrons—the realm of chemistry. But electrons and quarks do not have electrons as subsystems and so cannot couple with them—they do not have valence.

The valence interaction is what is called an "emergent property" of atoms—it only happens when electrons, protons and neutrons have assembled into atoms.

Valence coupling with electrons, then, is an emergent interaction of atoms while electromagnetic coupling with photons is an inherited one.

Bottom of hierarchy

Where does this hierarchical structure root itself—if a photon is a system, then from the above it must have its interacting subsystems, or coupling sub-subsystems. The suggestion in Superstring theory is that particles are self-sustaining vibrations, or solitons, in curled-up multi-exotic dimensions, and such "strings are not, of course, visible … impossible to detect by any means known to science today; they are mathematical curves."[1]

Is this pundit saying that the coupling sub-subsystems of particles are of the same stuff as mathematics? Perhaps not, but we really have not come across another suggestion.

While this rooting of the material hierarchy in such abstract stuff seems to be verging on metaphysics, it would tie up one loose end: if the root systems of the material hierarchy are really the same as—or even just similar to—"mathematical curves," then Wigner's "unreasonable effectiveness of mathematics" will no longer seem remarkable or needing of any further explanation.

Currently, the simplest suggestion of what our universe started off as is a 11-dimensional featureless sphere made of whatever it is that dimensions are made of.

The first thing of note that happened was the inflationary era in which the four time-space dimensions part company with the others and the four forces differentiate out from each other. The best mathematical description of this is currently group theory: "Towards the end of the last century, many physicists felt that the mathematical description of physics was

---

1 K. C. Cole, "A Theory of Everything,," The New York Times Magazine, Oct. 18, 1987, p.22.

getting ever more complicated. Instead the mathematics involved has become ever more abstract, rather than more complicated. The mind of God appears to be abstract but not complicated. he also appears to like group theory."[1]

The next phase is the conversion of inflationary energy into particle-pairs and the era of particle interactions which is currently described by complex numbers, Hilbert spaces, etc.

This is about as far as the "hard" sciences get (fully described mathematically) but the same principle applies. In a similar way, all the following developments—atoms, molecules, ... bacteria, ...primates etc. of the hierarchical structure—are all a result of the function working on the previous level.

We have already encountered the concept in Superstring Theory that the "stuff" out of which the "fundamental particles are made is more mathematical stuff than material stuff.

"This mathematical stuff is then processed by the natural law function such that "the entire sequence of events that unfold ...—the stars, the planets, the molecules, and the 'people'—are all just mathematical states ... a vast web of mathematical deductions spanning out from the starting state.... This speculative line of reasoning turns the Platonic position inside out. We no longer need to think of mathematical entities as abstractions that our material minds are battling to make contact with in some peculiar way. We exist in the Platonic realm itself."[2]

Made of Math: Run by Math: Described by Math. No wonder math has been called the Queen of the Sciences.

Coupling and forces

Before we move on we will mention here why classical physics does not describe interaction as exchange of subsystems but rather as forces, acting at a distance, that bodily move things around—obvious examples being the gravitational force, the electric force and the magnetic force. This is understandable when we realize that we can expect there to be consequences when systems couple with each by exchanging bits of themselves.

This is quite apparent in contemporary understanding of why the exchange of virtual photons in the electromagnetic interaction creates an apparent and measurable electric or magnetic force that moves things around. The explanation is quite simple—virtual particles can carry momentum along with them as they couple, and momentum—that key mix of: gravity and inertia—determines how mass moves through space. A change in momentum is a change in the way the mass of the system moves through space, it appears to be moved around by "forces" (in the classical sense).

Photons have momentum so that particles emitting and absorbing virtual photons will experience a change in momentum.

An electron coupling to another electron with photons can exchange momentum—its mass-through-space—in such a way that the change is such that the electrons move away from each other—there is a repulsive "force" between them.

It is this exchange of momentum via the virtual photons and the resulting effect on the history of the electron that is the classically-described "electric force" acting at a distance between charged particles.

This situation of subsystems being exchanged carrying their capacity for interaction is clearly a general one.

The other fundamental "forces" of boson coupling exert their influence on the fermion-bits-of-matter in a similar way

Gluons have the coupling capacities of momentum, spin, electric and color charge and transfer these from quark to quark. Unlike the other bosons, gluons have a strong tendency to emit and absorb gluons themselves—they couple strongly to themselves. This is just one of the reasons why the strong force is so difficult to describe mathematically.

W-bosons are like photons in that they carry momentum, they can also carry charge along.

---

1 A. Zee, Fearful Symmetry: The Search for Beauty in Modern Physics, Macmillan, NY (1986), p. 132.
2 John D. Barrow, Pi in the Sky: Counting, Thinking and Being, Oxford U. Press, Oxford (1992), p. 282.

An electron coupling the valence interaction between atoms, for instance, carries along with it its capacity to couple photons, its charge. But atoms inherit their ability to couple photons from the electron—charge is a secondary, inherited interaction in atoms and such a transfer of interaction capacity will clearly alter its interactions with other systems.

Similarly, a molecule coupling with a H atom in the H-bond—a chemical "force"—can expect that valance capacity is going to be transferred along with the H atom.

TYPES OF COUPLING

Every system interacts in some way—the neutrino, the helium atom and the putative Dark Matter albeit rather minimally—for even if there were such a thing as a system that did not interact, we would have absolutely no way of ever knowing anything about it. Hedging just a little, then, we can categorically state that all known systems interact—they have a tendency to couple with at least some of their subsystems.

In this section we will see how this tendency of a system to couple with its subsystems is, like its overall form derived from a wavefunction—the internal aspect of interaction. The external density of interaction will be derived from this by the hopefully-by-now-familiar random quantum operator assuming there is sufficient time available so that the random choice operator can be ignored.

The primary interactions of the system are not inherited from the constituent subsystems. All the other interactions—the secondary interactions—are. The valence interaction of atoms with is a primary interaction while the electromagnetic ability is "inherited" from the electron and quark subsystems. Only the valence interaction is novel to the atomic level, the electromagnetic and gravitational capacities are inherited from the electron's and proton's charge and mass. In the discussion we need only consider the primary interactions of a system—the interaction capacity it does not inherit from its subsystems. To describe secondary interactions later on all we will need is a frame shift.

The tendency of a system to couple with its subsystems is a reflection of the tendency of some, if not all, of its subsystems to disassociate from the system in some way—the subsystems are not monolithically integrated but are somewhat loosely associated. Another way of saying this is that there is a tendency for such a subsystem to "escape" from one system and gets "captured" by another system in some way. These labels are from the subsystem's point-of-view but it is all relative; the system's frame of reference these migrations are emission and absorption, they are coupling.

While all interaction wavefunctions basically the same, for purposes of exposition we have three possibilities for how two systems might couple with their subsystems. In practice, many interactions are a mix of them as they lie on a spectrum ranging from sharing through exchange to at-a-distance.

The simplest situation is that of exchange—the center of the spectrum. Crudely put, the wavefunctions of the two systems come into contact in some way and a subsystem hops from one system to the other. An example of this is the formation of sodium chloride, common salt, from sodium and chlorine atoms.

The exchange interaction occurs when the systems are in contact. As its name implies, interaction-at-a-distance, involves separation between the two systems. Here the subsystem hops out of the system—as in exchange—but then has to make it across the separation before it has the chance to hop into the other system and consummate the coupling. Our illustration of this will be the four fundamental interaction of physics in terms of charge—tendency to emit and absorb—and fields—the probability of making it across the separation.

The third possibility for coupling is the most interesting in its implications for it leads to stable structures, to links between systems, to the chemistry of atoms. The sharing wavefunction leads to subsystems being a part of two systems—or more—at the same time—the two systems are stuck together by a bond. The other interactions do not involve such implicit commitment. Our example of this will be the covalent chemical bond that links atoms into the molecules of life. An impressive example of this is a DNA molecule in which billions of atoms are linked by covalent bonds into a single, stable structure.

Clearly these categories are not that distinct: exchange interaction blends into indirect interaction as the separation increases, and into sharing in the other direction with the intimacy of sharing in a bond. Each of these paths of the subsystem that leads to coupling will have a probability amplitude, a little arrow pointing in an internal direction.

In the following we shall show that all three ways of interacting involve a correlation wavefunction, a constructive interference between one system's tendency to take in and the other system to give out subsystems.

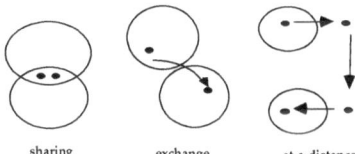

sharing   exchange   at-a-distance

The remainder of this chapter is devoted to describing the correlation wavefunction for each of these three varieties of coupling—they are basically very similar. Once we understand the correlation wavefunction, the rest is simple. The familiar step of a wavefunction becoming an actual density.

Always allowing sufficient time for the law of large numbers to counteract the unpredictable random aspect, the actual density of the coupling will be that of a probability density derived from the collapsed correlation wavefunction.

This density-of-coupling over time is the intensity of the interaction. We will return to this point in the next chapter when we look at the consequence of interaction—the higher the intensity, the more are the consequences. But first the internal aspect of interaction in the perspective of the new physics.

Valence

In our discussion of form we restrained the discussion to that of isolated, stable systems that was not involved in gaining or loosing primary subsystems.

The isolated, lithium atom, for instance, has no tendency to lose its outer, solitary 2s electron. Such a lithium atom is, however, not a happy one in the sense of being in a state of least resistance. There are two things that are paths of high resistance: for the system—the electronic state is not that of a noble gas, and the outer electron is not paired. We have already seen how little arrows explain why the pairing of electron spins in an orbital is a high-probability state. The noble gas state is similar in that it is the state where all the orbitals in a shell—the main quantum number, n,—are filled with paired electrons. This is such a low resistance state that almost all the chemistry of atoms can be explained by the impulse to inhabit this blissful—I mean, low resistance—state.

The coupling capacity of atoms is very significant in chemistry. The coupling capacity of an atom for electron exchange called its electro-valence. The overall tendency to take in an electron is called the electropositive character of an atom while the overall tendency to give one out is its electronegative character. An atom is usually characterized by which of these tendencies is the stronger though some, like hydrogen are equally capable in both directions.

The coupling capacity is measured by interacting atoms together to give a relative measure of such tendencies. Thus a current definition: "Electronegativity is the relative tendency of an atom to acquire negative charge…. [for example the] relative scale in which the most electronegative, fluorine, has a value of F: 4.0… are: O: 3.5, N: 3.0, C: 2.5, B: 2.0, Be: 1.5 and Li: 1.0."[1]

This is simple exchange. Exchange involves the correlation between the positive tendency of one system with the negative tendency of the other.

Not all atoms are so eager to participate in exchange interactions. If both positive and negative coupling capacities are zero the system has no tendency to lose or gain a subsystem. This is the situation for an interaction-indifferent system such as the helium atom.

---

1 D. M. Considine (1983) ed. Van Nostrand's Scientific Encyclopedia Van Nostrand, NY (1983) p. 1067.

Nuclear forces

Another basic example of interaction in contact is that of the strong force that holds the atomic nucleus together. The protons and neutron exchange virtual pions when very close to each other—a derivative of the strong color force that holds the quarks together inside the nucleon. The consequence of this is a massive transfer of momentum that pull the nucleons together with a fierce force—the quark degeneracy pressure making sure they don't get too close. It is this attraction that holds the nucleus together. It has to be strong because the positive protons that are right on top of each other have an intense electromagnetic repulsion that has to be overcome.

It is a balance between the pion exchange pulling the nucleons together and the photon exchange pushing them apart. The balance is such that two protons will not stick together by themselves—there is no helium nucleus with just two protons. This is just as well, actually, for if not so all the hydrogen atoms—single protons—in the sun would rapidly combine and its 10 billion years worth of energy would be released rapidly in a titanic explosion that would wipe out the solar neighborhood.

With just one neutron added to the mix, however, the balance is radically shifted—a helium-three nucleus—one neutron and the two protons—is a very low resistance state—energy is given off in its formation from the free nucleons. The neutron indulges avidly in the attractive pion coupling but not in the repulsive photon coupling. In the sun the only way two protons can stick together is if one of them changes into a neutron first. Then they can embrace with pion coupling—no disruptive photon coupling—with great release of energy. This is hydrogen-2, the deuteron, the first stage in the nuclear burning of hydrogen in the sun. The trick is getting a proton to change into a neutron—the reverse of neutron decay—and this involves the weak force. Being weak, it takes billions of years, on average, to flip a proton into a neutron and thus the essentially-slow rate of burning at the center of the sun.

The pion coupling does have one limitation—it depends on the virtual pions. And pions are quite massive—about a half the mass of a proton. Such a massive virtual particle has a very short lifetime—such disobedience of the law of the conservation of energy cannot last long enough to create a quantum of action. So brief is the allowed lifetime of these virtual pions that, even moving close to the speed of light, they can only cross distances about the size of the nucleon. It is a very short range force—even though it is very strong, its influence is severely limited to the size of the nucleon. This is why nucleons separated by more than their diameter do not attract each other by pion exchange. This is also why the weak force is weak, its victual particle is super massive and has a correspondingly tiny sphere of influence.

The repulsive photon exchange, however, has no such limitations. The virtual photons have zero mass which gives them infinite range. Thus in a massive nucleus a protons is only attracted by the nucleons in its immediate vicinity while it is being repelled by all the other protons in the nucleus. Eventually this accumulative repulsion overwhelms the non-accumulative strong force and the nucleus becomes unstable. By the time we get to uranium with 96 protons squished in the tiny nucleus the balance swings over to the repulsion and the nucleus is unstable, it tends to break up, it is radioactive.

Sharing

We now have a basic picture of interaction by exchange. Next we will look at interaction by sharing. The two are very similar in that sharing can be thought of as partial exchange,

The simplest example of this is the hydrogen atom. It is in a doubly high-resistance state—it has a singleton electron and it is one short of the desirable helium-like state. There is a certain tendency to lose an electron and a similar tendency to take one in. This equal matching of tendencies precludes either one of two hydrogen atoms gaining total control of both electron. Rather, the constructive interference between the correlations creates correlation wavefunction in the tendencies in either directions are the same. The correlation wavefunction is just like that for the exchange except that all four coupling capacities appear—not just a plus of one and the minus of the other.

Both directions are important. From this QPF comes the actual density of the coupling, the probability density, or intensity, of sharing.

This is all the theory we need to understand the nature of the covalent chemical bond.

BONDS

When two hydrogen atoms share their electrons the two high-resistance states disappear. There are no longer two unpaired electrons, there is a single, low-resistance pair. And both atoms can now lay claim to the helium-like structure—they have a filled main orbital even if its a shared one. One electron from each atom inhabits the bonding orbital. This is such a low-resistance thing to do that this sharing holds the two atoms together in a covalent bond. Such a pair of electrons inhabiting a shared orbital is usually symbolized by a single line joining the atoms.

A simple picture of this bond is that it is a resonance—there are two exchange interactions going on at the same time: each hydrogen atom has the electron pair 50% of the time. This picture makes apparent the involvement of all four coupling capacities in the correlation. The bond is a resonance of two forms where they "alternate" being in the helium-like state of being by having the pair of electrons. There should be no problem, at this point, in understanding how a wavefunction can be a mix of opposite states. Each system being helium-like 50% of the time has lower resistance than both of them being singletons 100% of the time. Being 50-50, the bond is not at all polarized—the hydrogen molecule does not, for instance, participate in the hydrogen bond

Resonance is commonplace in chemistry—many bonds are best understood as resonances of more elementary wavefunctions.

In chemistry, the correlation between two 1s orbitals are hybrid "molecular orbitals." It is the filling in of these by electrons that is the covalent chemical bond.

There are two ways in which the 1s orbitals hybridize—one low resistance, the other not.

The bonding orbital is the low-resistance state for a paired set of electrons. The anti-bonding orbital is a high-resistance state for one, let alone two, electrons to be in.

Helium molecule—not

Exactly the same thing holds for two helium atoms, each with two electrons in the 1s orbital. When they are in proximity, the sharing orbitals can be filled up.
In such a case, however, while two electrons pair up in the bonding orbital, the other two would have to inhabit the anti-bonding orbital.

This is such a high-resistance state of affairs that helium atoms do not form a chemical bond with each other and helium molecules never form. Helium atoms are so self-satisfied that do not like to get to close together and, consequently, helium gas only reluctantly turns into a liquid when the temperature is almost absolute zero so they have no kinetic energy to get away from each.

In the terms we have just established, helium has a zero coupling capacity for valence, both plus and minus components are zero. So the correlation is also zero.

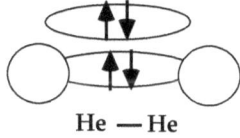

Carbon bonds

By far the most significant example of such equal sharing is the bonding ability of carbon. We have already discussed how the carbon atom is exactly half-full of the electrons it needs to complete its 2 shell. The s-orbital and three p-orbitals hybridize into four equivalent SP3 orbitals, one at each corner of a tetrahedron.

When two carbons are in proximity two hybrid orbital open up—a bonding and an anti-bonding orbital. Each carbon atom contributes one electron to the pair, the carbon-carbon bond that is, without exaggeration, the basis upon which life is built.

In our simple description with quantum operators, the bond is a filled in wavefunction with probability density.

This is the single carbon bond. The other three orbitals make bonds in exactly the same way. This is such a satisfying state of affairs—such a low-resistance state—that carbon is very reluctant to break this bond. It is this reluctance that makes diamond—each crystal a single molecule in which every carbon atom is singly-bonded to four others.

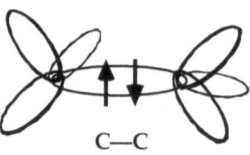

Carbon likes its company so much that it will form double bonds—the unsaturated in fat—and even, on occasion, triple bonds. Making four bonds, however, is distorting things to much and are anti-bonding.

Carbon also bonds well by equally sharing with hydrogen, and just about every element for that matter.

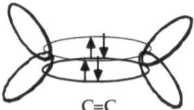

The singleton 1s orbital of a hydrogen hybridizes with a SP3 of the carbon—their correlation wavefunction—and each contributes an electron to the low-resistance pair that inhabit the bonding orbital. Four such bonds satisfy both the hydrogen and the carbon in the methane molecule.

The measure of the sharing tendency of atoms—its valence—is somewhat different to the electrovalence for complete exchange. For instance, fluorine, which has a distinct tendency to rip electrons from others is tamed by carbon's sharing tendency to such an extent that the fluorocarbons are amongst the most stable of molecules, famous for their non-stick aspect so self-satisfied are they.

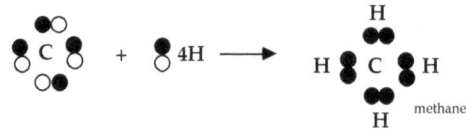

Unequal sharing

Where different atoms are concerned we do not always see such fair sharing. For instance, the bond between oxygen and hydrogen in the water molecule. Both atoms contribute one electron each to filling in the correlation. We can think of the bond as a resonant form where the oxygen has both pairs 60% of the time—which puts it into the desirable argon-like configuration—and the hydrogens have a pair each just 40% of the time putting it in the helium-like state.

As the electrons spend more time with the oxygen it has a relative negative charge compared to the hydrogens, and this is the basis for the all-important hydrogen bind. The polarity indicates that the positive and negative coupling capacities for valence are not equal—that the tendency for the oxygen to take in is greater than the tendency of the hydrogen to take in. It is this polarity that accounts for the ability to hydrogen bond.

There is a distinct probability of a molecule of water splitting into a hydrogen ion and a hydroxyl ion. Although the hydrogen ion is only a bare proton, it does not behave as an "elementary" particle in physics, rather it behaves as an atom with all its orbitals empty. All this bare atom needs to be helium,-like, however, is a pair of electrons. It finds these by latching onto another oxygen on a neighboring water molecule—in the sharing the oxygen contributes one its electron pairs not already involved in bonds. These outer pairs are called lone pairs.—nicely filled orbitals with nothing blocking access to them. The three hydrogen bonds to the oxygen are all equivalent—the positive charge gets smeared out over the three hydrogens—a quite low-resistance state.

The hydrogen ion hybridizes its empty orbitals with those of the lone pair orbitals on the oxygen. A filled bonding orbital results, binding the hydrogen ion into a hydroxonium ion. As water molecules are everywhere, this is how the hydrogen ion always is in solution. Here we see the significance of empty orbitals—they are just as real as the inhabited ones.

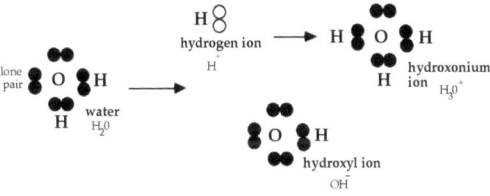

The proton can easily skip from one water molecule to another and this is how the hydrogen ion moves through liquid water. Directed and controlled, such proton transport is of fundamental importance to living organisms where "proton pumps" generate almost all the useful energy for a cell.

It is the sharing interaction that leads to linkage of systems into stable configurations. Such interaction is clearly going to play a significant role in subsystems hooking up as stable systems and systems hooking up together as supersystems. We saw this example where the consequence of the sharing interaction of hydrogen atoms and oxygen atoms is a water molecule. This molecule can interact in ways the atoms cannot—it can H-bond for a start. A system higher in the hierarchy of coupling capacities has emerged from systems lower in the hierarchy.

Where the sharing interaction can be expected to be pre-eminent in big-picture system building, it is not alone. We see an example of system-building by exchange at-a-distance where electrons and protons electromagnetically interact. As the electrons fill in the orbitals provided by the nucleus; an atom emerges and valence is on the scene. Natural law will determine the path of least resistance for valence and it will contribute to the internal wavefunction; this will determine the probability of what the atom will do. in the usual way

At-a-distance Interaction

To conclude this chapter we will look at interaction at a distance where separation between systems is involved.

The process has three basic steps from the subsystem's point of view: It escapes from one system. It then travels as a free system. It is captured in the vicinity of the another system.

From the systems' point of view the process is similar. One of them emits a subsystem. The subsystem makes it across the separation. The subsystem is absorbed by the second system. The first and third steps in coupling at a distance involve subsystems leaving and entering a system. This is very similar to the coupling capacity we have already established for the exchange interaction; except hat here the giving out and taking in does not depend on there being another system in the vicinity.

Charge

Coupling at a distance can only occur if one system has a tendency to loose one of its subsystems and the other has a tendency to gain it. There is a positive coupling capacity to give out a subsystem and a negative coupling capacity to take one in. Our example of interaction-at-a-distance is the electromagnetic—the coupling with photons. A virtual photon in the quantum-foam structure of the electron has a probability amplitude to be emitted with little consequence for the structure and stability of the electron. In the new physics, it is this probability of gaining or losing a photon that is the measure of the electric charge off the electron.

This probability of emitting and coupling subsystems is called the "charge" of a system in basic physics. The symmetry of natural laws—acting on the internal aspect in the new physics—ensures that these two tendencies will be equal—at least in the sense of being conjugates of each other. When natural law is providing the wavefunctions, the plus and minus directions are always equal in probability. Note the caveat "when natural law is providing," for later we will deal with situations where natural law is not the direct provider of the wavefunction and the two directions are no longer necessarily symmetrical.

In basic physics, the probability of a system emitting and absorbing a coupling subsystem is called its charge. The well-characterized electromagnetic interaction is coupling with virtual photons. The probability of a particle emitting or absorbing a virtual photon is called its electric charge and is measured by what is called the fine structure constant or coupling constant. For a with unit charge this probability is given by the collapse of the coupling capacity. The actual density will equal the probability density over time.

One can think of this in a simple way: given 137 opportunities to emit or absorb a photon, the electron will do it once.

The reason why the probability is the same for the positive and the negative directions is the reversibility of the natural laws that govern the internal realm. In fact, as the amount of energy tied up in each individual coupling event with a virtual photon is very small, its extension in time is so fuzzy and impossible to pin down that it is impossible to say which charged particle give out the photon and which one did the taking in—all that is certain is that the exchange did take place.

As we are really quite ignorant of the inner structure of the electron, the fact that the fine structure constant is this has to be "added by hand" into current theories as it cannot, as yet, be calculated from first principles.

It is this number that is the proper measure of electric "charge" in modern physics. Incidentally, the same number measures the magnetic "charge" as well which, as it turns out, is simply another consequence of things not routinely traveling at the speed of light. Traveling at low speed, we interact with virtual photons in two seemingly distinct ways. The description of this effect is obtained by combining Maxwell's classical field equations with Einstein's relativistic ones.

Nuclear forces

The weak force is remarkably similar to the electromagnetic—so much so that they are often referred to collectively as the electro-weak force. The probability to absorb or emit a weak boson is exactly that as for the photon. The big difference is that the weak bosons are massive and cannot get very far—the mobility playing a determining role here. The color interaction of quarks is much "stronger"—hence its name—than the electro-weak interaction.

This is because the coupling constant for gluons—the probability that a quark will absorb or emit a gluon is essentially unity:

Given the opportunity to emit or absorb a gluon, the quark will always do so.

To make things even more complex, gluons themselves have color charge, they also couple with gluons—unlike the photons which have zero tendency to absorb or emit other photons. And these sub-coupling gluons, o to peak, also couple with each other. It all gets very messy as all of this has to be taken into account to "solve" the equations that we describe the color interaction. It recently took an IBM supercomputer almost year to process all the terms that have to be taken into account just to figure out the interaction of the three quarks making up a nucleon. Apparently, almost 25% of the "mass" of a proton is actually the energy tied up in gluons coupling with each other. In comparison, the "mass-energy" of the electromagnetic field coupling the proton and electron in the hydrogen atom can be ignored for all but the mot accurate of computations.

To even things up a little, however, because of all this promiscuous coupling, gluons don't get very far; just over distances commensurate with the size of a nucleon. Both gravity and electromagnetism do not suffer from this limitations of scale

Gravity

On the largest of scales, electromagnetism's tendency to cancel out its effects—because of the overall balance of positive and negative charges in nature—leave the largest of scales to be ruled by the unimaginably-weaker force of gravity; graviton coupling.

The classic illustration of this disparity in strengths is that the gravitational force of the proton in the hydrogen atom on the atomic electron is equal to the electromagnetic force of a proton on an electron at a distance of a star in our neighborhood, about 100,000,000,000,000 miles away.

On the largest of scales, of course, even these minuscule forces start to amount to something. Gravity rules by dint of the absence of a negative type of mass/energy that could cancel out the attraction of regular mass/energy for itself. It is gravity that rules the structure of planets, stars, and so on up to superclusters and the cosmic level. The "gravity charge" or mass of a system is based on the probability it will absorb or emit a graviton:

Fields

We now have a measure—in the coupling capacity—for the first and last step in the three stages involved in interaction at a distance:

a system looses a subsystem,

the subsystem moves,

the subsystem is taken in by another system.

Now we will deal with the intermediate step, the mobility of the coupling subsystem.

While step one and three involve the systems, the second step does not. The freed subsystem is now an independent system. As established, the such a system has a probable future, and the autonomy of the system will pick one of these. This is an open-ended history and it will involve an open-ended wavefunction as discussed earlier.

All the other paths are those infamous not-taken ones.

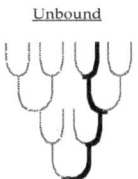

This wavefunction gives the probability of finding the system at a particular location as the systems moves through it exhibiting its random nature. This spread out wavefunction is called a field. If a lot of systems are involved, the field is the cumulative probability of all of them. So, for example, the overall probability of finding a virtual photon at a location is called the value of the electromagnetic field at that point. If just one system is involved, the random choice operator will have to be taken unto account but, when large numbers are involved, it can be ignored.

The electromagnetic field always involves lots of photons so we can ignore the random aspect.

Naturally, we could measure this probability density by adding up all the little arrows, a tedious and time-consuming method, but one that gives the correct answer. A much simpler method, and the one used throughout physics, is the use of a field equation. This is similar to the way that the Schrödinger equation simplified calculating the electron orbitals of hydrogen. In fact, Schrödinger's equation is a field equation, one that treats the electron field. In the form of systems, the fields dealt with the probability density of structural subsystems; for interaction, the fields deal with the probability density of coupling subsystems.

So much so of fundamental physics can be expressed as field equations, in fact, that some physicists have gone so far to declare that objective reality is fields, and field alone. We have not taken this route, in our point of view the fundamental reality is systems of interacting subsystems. Mathematically, however, they are equivalent.

A field equation simply allows one to calculates the density of coupling subsystems at any point one is interested in. For instance, they are capable of calculating the quantity we were just discussing, the probability density of the mobile subsystems making it from system 1 to system 2.

In our perspective we can say that field equations give a measure of the probability density of the coupling subsystems at any location.

All the field equations of modern physic—and they are daunting in their details—can be thought of in this way. A simple way of thinking about this is that the system "throws" out this field of influence based on its ability to couple subsystems. One point too note is that, echoing the way that mathematicians recognize the "null set" or set-without-members as a significant entity, physicists accept that the field is still there even when its value is zero. The definition of the vacuum is that all fields have a zero value

The field is theoretically measurable by a "minimal test particle," a particle that, while it couples with the field, does not itself alter the field in any way. To measure the field, the test particle is placed at a location and the amount of coupling is noted, a measure of the field at that location. (Such a measure will be a measure of the consequence of the coupling—such as a force—the topic of the next chapter.)

Travel

The key to interaction is the probability that the coupling subsystem will make it across the separation between the two interacting systems. As always, this probability is given by the collapse of a wavefunction, in this case, the field.

Interaction at a distance has three steps: emission, travel across, and absorption. The wavefunctions for these three steps are the positive capacity of the first system, the field of the coupling subsystem, and the negative capacity of the second system. The correlation between the two systems will be the constructive interference between these three. The intensity of the interaction will be the collapse of this correlation wavefunction.

This is a somewhat hybrid expression as it combines attributes of bound subsystems—the positive and negative coupling capacities of the system—with attributes of unbound subsystems—a QPF, the field wavefunction.

Exactly the same holds for coupling in the opposite direction. For all the fundamental; interactions where the tendency to emit is the same as the tendency to absorb, the correlation will involve both directions.

Electromagnetic field

A good example of a very successful field theory is the electromagnetic influence of charge at a distance. The electromagnetic interaction is carried, as the physicists say, by virtual photons. The classical field equations of electromagnetism give the probability density of the virtual photons at a distance from the "charged" particle. The entirety of this over all space is the "electromagnetic field" generated by the particle. As mentioned, the electromagnetic influence spreads far indeed in that the electromagnetic influence of an electron and proton separated by interstellar distances equates with their gravitational influence at atomic distances.

The electromagnetic field at a location is nothing more than the probability of finding virtual photons there to couple with.

In our general discussion of emission we spoke of a subsystems escaping from the system This process is not understood. Just what happens at the start of a photon's journey—or at its end—is not understood. But leave and enter they do; photons begin and end on electrons. The situation is made even more hazy by the relative spatial extension of an electron—which does the absorbing and emitting—and a photon. While the spatial extension of an electron is less one millionth of a nanometer, the spatial extension of visible light is huge in comparison on the order of millions of nanometers.

Whatever happens, the initially-released virtual photons take of at the speed of light—they spread out symmetrically—they have no preferred direction—and expand into space. The field equations describe this very simply, the electromagnetic field, the density of coupling subsystems, falls off as the square of the distance.

The field equations do not take their inspiration from sound waves in organ pipes, rather they are modeled on the density of fluid flow.

Our example of indirect coupling will be the electromagnetic interaction of the electron and proton which is basically:

1. the electron (or proton) emits a virtual photon—its charge
2. the photon travels from place to place—the electromagnetic field
3. the proton (or electron) absorbs it—its charge

The discussion is applicable to all four fundamental interactions as they are all similar though the terms used are somewhat varied.

Interaction	emit/absorb	"charge"	subsystem movement
electromagnetic	γ	electric	electromagnetic field
gravity	gr	mass	spacetime curvature
weak	W, Z	weak	weak boson current
strong	gl	color	gluon field

Getting across

Unless the "background" over which coupling-at-a-distance occurs is very inert, it can have a great influence on the probability that a subsystem will make it from one system to another.

It could be absorbed and never make it, for example, or be retarded by being absorbed and then emitted along the way.

In the electromagnetic interaction coupled by virtual photons, for instance, the measure of how they are influenced by the intervening space the photons are traversing is called the dielectric constant, a measure of the ability of the virtual photons to traverse whatever it is that separates the interacting systems. Some systems, like iron atoms, enhance the mobility factor in the electromagnetic interaction but even the "nothingness" of the quantum vacuum foam has a slight, and measurable, effect in retarding photons as they pas by.

As mentioned, it is the mobility of the coupling subsystems, rather than the tendency to couple, that gives the weak interaction its moniker: "…the amplitude for a particle to emit a W is really no smaller than the amplitude for the particle to emit a photon, but the W is so massive that the probability amplitude for it to pass from one particle to another is very small—it gets so 'tired' that it's prone to turn right back. This [explains] why the weak interaction is so much weaker [than the electromagnetic one]."[1]

Inflation

On the other hand, as far as we are aware, nothing seems to influence the mobility of gravitons, the gravitational interaction is oblivious to what lies in between systems. In the very early history of the universe—about 10^{-34} of a second[2] after the Big Bang—it is widely held that an exponential expansion occurred, an abrupt inflation of atomic-size extension to galactic supercluster dimensions. When this inflation abruptly stops, the shock energy of this change then kicks off the "classical" hot Big Bang about a trillionth, trillionth of a second after the true beginning point.

The inflation is driven by a cosmic negative pressure field which is like negative gravity—it is intensely repulsive in its effects. Conditions were such that graviton coupling was powerful and expansive, unlike its pale descendant today which is feeble and contractile.

In many, if not most, cases, the subsystem mobility is such that the interaction decreases with increasing distance, usually described by an inverse square-of-the-distance law which simply reflects the geometric realities of volume with distance.

This is not always so, however; the interaction of quarks via gluons is at a minimum when they are close together but rapidly increases in intensity as they move apart—the 'infrared slavery' that further complicates our ability to fully describe color charge. It can also be very complex, as it is in cells coupling with hormones and other factors where the transportation by blood is involved.

---

1 Anthony Zee Fearful Symmetry Macmillan, NY (1986) p. 221

2 While this sounds like a short period of time—and it is—we note that this is still well over a billion quantum "ticks" of the Plank Time, plenty f time for things to happen in!

Hydrogen bond field

Field formulations are a useful perspective for more complex at-a-distance interactions. For instance, it is useful to think of hydrogen boding in terms of fields. The exemplar of this capacity is water in bulk. The electromagnetic field is the probability of finding a virtual photon at a location; the H-bond field is the probability of the orientation of a water molecule at a location. water tends to structure the water around it, to attain the low-resistance state of the ice-like mesh The molecule structures the water around it, and this field can stretch an appreciable distance before it is overcome by random thermal motion or the imposition of the field of another. Water molecules are equally matched, they move each other around.

This equality does not hold when massive molecules are involved. Many molecules with oxygen (and nitrogen) in them are good at hydrogen bonding as are almost all of the molecules of life. As they are massive they move the water around much more than the water moves the massive molecule around. The molecule imposes its H-bond field on the surrounding water. In the formation of macromolecules, however, the cumulative push and shoving of many water molecules H-bond fields is very significant in moving the molecule around. An example is the spontaneous folding of an amino-acid chain into an active protein enzyme, a process driven by the combined desire of the macromolecule and the multitude of water molecules to structure into a state of least resistance. The process clearly involves wavefunctions with steep gradients in them for a "denatured" protein can refold into the active form in milliseconds.

This attempt by the molecule to structure the water around it will impinge upon the attempts of other molecules to structure the water around themselves. This would be coupling through the H-bond field. The form of biological molecules in water is not just that of the atoms it is composed of, it also includes the structure it imposes on surrounding water. The surrounding water molecules have to be included in the structure of the molecule.

Water is a somewhat polar molecule. While the bonds between the hydrogens and the oxygen atoms are predominantly sharing they also have quite a bit the nature of exchange as well. The oxygen takes more than its fair share, it pulls the electron pair it shares with the hydrogens close to it, making it relatively negatively charged, leaving the hydrogen somewhat positively charged. In comparison, the bond between carbon and hydrogen is scrupulously fair and there is zero polarity and thus no hydrogen bonding. The negative oxygen of one water molecule can attract the positive hydrogen of another molecule, this is the hydrogen bond.

These bonds are directional and the molecules have a sticky tendency to mesh with each other. When thermal energy is low the stickiness of these bonds is sufficient to hold the molecules in place and we get the open mesh structure of ice—it floats because of this open mesh structure.

Steam results when the thermal energy is much greater than the stickiness and the molecules fly free of any bonding. Between the two is the magic zone that allows for life. When the thermal energy is similar to the stickiness energy the alignments are temporary—they form and are then disrupted—and we have liquid water. There is alignment as in ice but it is only temporary as thermal motion tends to break it apart. Keeping in mind that we are really speaking about paths of least resistance, we can crudely characterize the hydrogen bond as the "desire" of water to take up the ice mesh structure in the same way that chemistry can be crudely described as the "desire" of atoms to take up the noble gas electronic configuration—filled paired shells.

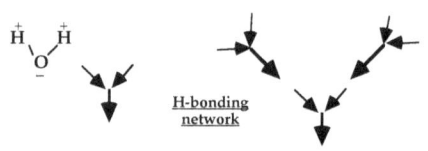

The tendency to hydrogen bond is carried outwards in the structuring and polarization of surrounding molecules. We can think of the water surrounding a biological molecule as a field of structured water—a wave of alignment—and all the interesting interference effects that can be described by complex numbers. Just as the electromagnetic field is a simple description of the probability of absorbing a virtual photon at each location so the hydrogen bond field is a simple description of the probability of a water molecule having a particular configuration at each location.

Change in history

We will now deal with the simplest, and most common, consequence of interaction where one system influences the history of another system. We are still basically restricting the discussion to peer interactions—systems interacting with systems on the same level in a hierarchy. Two systems influence each other's history as a general consequences of the fact that subsystems take their capacities along with them as they change allegiances during the interaction. In the most general sense, the consequences will depend on how much coupling capacity is carried along by each subsystem and how many subsystems are being coupled.

The consequences will be proportional to the intensity of the interaction—the collapsed correlation wavefunction—and to the coupling capacity carried along by each subsystem.

The amount of consequences will depend on the intensity of the interaction—the amount each system carries along with it times the number of them making the trip. These consequences of interaction can be roughly equated with the forces that appear in classical science descriptions.

At this point the discussion bites its own tail, so to speak. Very much earlier we spoke of the source of the probability amplitudes that have informed our discussion of modern physics. We spoke of natural laws described by action equations. The action equations take into account all the contributions of each interaction. The item that actually appears in the equation is what we have been calling the consequence of interaction. When this changes, the wavefunction changes. The system now has a new wavefunction with a new collapsed probability density. The system will follow one of the probable histories in this new set-up subject to the vagaries of the random choice operator, The sequence describing simple change is: 1. internal correlation 2, external filling in 3. transfer of coupling capacity 4. change in wavefunction 5. change in history.

Leaving out all the system labels for simplicity's sake we have a simple sequence of wavefunctions.

The subsystems carry their capacity to interact along with them on their travels. The capacity to interact that is being carried along by the transfer of subsystems are the secondary interactions of the system itself.

We established that a system has an overall capacity to interact, internal system, that was the composite of two qualitatively-different types of interactive capacity.

a. primary not inherited coupling with primary subsystems, unique to the system itself, not a capacity possessed by any of the system's subsystems

b. secondary inherited coupling with secondary subsystems, a capacity possessed by primary subsystems; includes the tertiary and on down as similar.

The capacity of the system for secondary interactions is inherited from the primary subsystems so when those subsystems are transferred they take the secondary interactions of the system along with them.

CONTINGENCY

The concept of contingent history that pops up throughout the sequence is just the simple requirement that there be interaction for there to be change If there is no interaction there is no change. But systems can only interact with each other if they are in the vicinity of each other (or at least close enough for coupling at a distance to be significant. For simple systems, in the vicinity can be equated with being close by each other. We are no longer talking probabilities here, the two systems have to be in the same place and the same time—a particular set of histories. As we have established, a particular course of history involves the random choice operator. The random choice operator of both systems must pick the same place at the same time—all interaction occurs in the present—so that they end up on the scene together—ripe, so to speak, for interaction. This is the contingent side of history and it very much involves randomness and is to be avoided if possible. And, as noted, possible permutations of even a small number of possibilities involve large numbers.

This would be an impossible situation if an infinite number of possibilities—a continuum—was involved as in classical physics. Luckily, the way wavefunctions interfere does not involve an uncountable infinity of states, not even a countable infinity of them, but just the combinations of a small set of small numbers. Even better, nature almost always involves large numbers of systems on the scene at the same time. Even tiny-probabilities can (relatively) quickly appear on the scene—in fact, sooner or later, any not-exactly-zero probability must appear on the scene. In this way, the influence of the random contingent aspect of history is somewhat nullified

Contingency does rule in the actual origin event, however, as the random operator comes into play to get them there. At some point in time the subsystems were on the scene, they did fill in the system wavefunction, an the system emerged on the scene.

Contingency also enters as the larger environment intrudes:. An example would be the results of a slit experiment performed when the nuclear pile next door goes critical and explodes. We will deal with interaction with the environment after we have dealt with two-system interactions.

This contingent history has an internal and external component: the systems have to be in the neighborhood and they have to have a significant correlation. The contingent prerequisite for interaction to occur is the systems must be in a situation—a configuration—such that there is a non-zero correlation between them:

internal correlation of systems

external configuration of systems

The consequence of interaction also has an internal and external aspect.

internal change in wavefunction and probable future

external contingent history actually followed

Movement

We will now look at examples of this somewhat general discussion. One example is the electromagnetic interaction where the exchanged virtual photons carry momentum—the gravitational interaction—along with them as they shuttle between the interacting systems. It is this exchange of momentum that is the electromagnetic force of classical science.

force of interaction (external consequences)

= intensity X amount carried by each

The capacity for coupling transferred by this flux of photons involves just one, the capacity to couple with gravitons. Early in the discussion we saw that graviton coupling had two aspects: gravitational mass, the ability to couple, and inertial mass, involving a change in coupling. Virtual photons do not transfer gravitational mass/energy—being virtual, this is to be expected. (Real photons, on the other hand, do transfer real energy.) Virtual photons do, however, transfer the inertial aspect of graviton coupling. This inertial aspect is measured by momentum, a measure that is as well-defined in quantum physics as it is in classical physics—unlike "energy" which, as we have seen, can be somewhat fuzzy over time. In classical mechanics, momentum is the product of mass times velocity:

When virtual photons are exchanged they transfer momentum between the coupling systems. The input of the electromagnetic interaction to the electron is the transfer of external photons and internal momentum.

Electromagnetic force

The transfer of momentum carried by the virtual photons is such that the electrons move apart, their inertia is altered by the coupling. The mass/energy of the electron, on the other hand, remains constant as the photons do not transfer it. It is this moving apart that classical physics calls the "electromagnetic force" pushing them apart. This bodily movement of the electron is external, and it is a reflection of what is happening on the internal, wavefunction of little arrows.

Momentum transfer is important at every level in the hierarchy of matter for almost all the higher capacities involve subsystems with real mass and real energy—so, unlike the virtual photons, they transfer mass and momentum along with them. Much of the movement of matter derives from this transfer.

The movement of the system, as a consequence of the coupling, can influence the correlation. Our example are two electrons interacting and, as noted, they move away from each other. As they separate the intensity of the interaction falls off. Less photons is less momentum transfer. Less momentum transfer decreases the "force" pushing them apart, the acceleration apart decreases with time. This is a negative feedback, the interaction, and hence its consequences, decreases over time.

Lipids, on the other hand have a positive feed back to the movement towards each other, the closer they get, the easier it is to displace discontented water molecules and the hasten together.

While virtual photons only carry momentum, real photons carry both energy and momentum, both of the aspects of graviton coupling. So a slow-moving electron that absorbs a high-energy gamma photon has both its momentum and energy changed, it becomes a high-energy electron zipping along at high speed.

When an atom absorbs a real photon, one of its electron moves to a higher energy state. Such "excited" atoms (or molecules) often have a quite different tendency to interact compared to their "ground" state. It is this phenomenon that underlies the photosynthetic powering of almost all life on earth: a photon-excited electron in a chlorophyll molecule is whisked away down a metabolic pathway; the energy in the ensuing charge separation is then used to power a cascade of chemical transformations that ultimately turns carbon dioxide and water into carbohydrate and oxygen.

Proteins

Unlike the electron and proton which basically only couple in one way, proteins have multiple ways of coupling. Proteins are remarkable for the versatility of their interactions and do most of the "doing" in a cell. Proteins, for one thing, are marvelous organic chemists and are capable of many chemical syntheses impossible for the man in the lab. It is truly remarkable what just twenty-odd amino-acids can do when they are linearly liked in their hundreds. All these interactions will contribute to the external and internal input to a protein.

Almost all of the important interactions of proteins involve the spatial pattern of the interactive capacities on the extended system. Some of these important "patches" of interactive capacity on a protein "surface" are: ± H-bond ordering, ± charge, lone pairs, empty orbitals, metal ion interactions, aromatic ring resonances, etc.

Lipids

The ordering about of water by H-bonding capacity is one of the main contributors to moving large molecules around into their active structures.. All of life's molecules are in an environment of water molecules and have to deal with water's determination to minimize its resistance by forming oriented fields of H-bonding. The interaction of a single water molecule with a macromolecule has consequences for both of them—by the reversibility of natural law these will be equal and opposite. The tiny water molecule is drastically altered while the huge macromolecule gets a tiny tug. But there are a lot of water molecules around and the tiny pushes and pulls can add up to significant imbalance which the macromolecules bodily moves to correct. The movement stops when all the pushes in one direction are balanced by the pushes in the other. This is just how a massive amino acid chain folds into its active form just from myriad tiny tugs of water molecules.

A molecule that in has no capacity to H-bond will have around it a shell of very unhappy (high resistance state) water molecules. A excellent of example of this is a lipid (fat) molecule that has, as its main bulk, a long hydrocarbon chain in which the hydrogen and carbon fairly shares their shared electron pair—the molecule is non-polar as it is the greedy tendency of the oxygen atom to hog the electrons that polarizes the water molecule and sets the stage for H bonding.

This shell of water molecules is highly imbalanced—on the lipid side each molecule is unable to form a H-bond while on the bulk water side it is H-bonding. This unequal state is surface tension and its consequence is repulsion. This is a strong "force"—a small bead of water will lift itself up against gravity as it beads on a waved surface. This time it is the lipid that is repelled—it moves away from the water and into itself. "Oil and water do not mix" is a significant principle in the structure of the macromolecules of life..

Unlike the multi-talented proteins, this is about it for lipid interactions except for a slight stickiness most molecules feel for each other—think Post-it-Notes—called the Van der Walls attraction. This is why as any gas cools it eventually turns into a liquid—the stickiness and the kinetic energy of motion are similar. This is residual electromagnetic force based on the fact that the negative electrons in the atom are so spread out compared to the point-like positive nucleus that perfect cancellation of charge is not possible, the positive charge is not totally shielded by the electrons even in the neutral state. Helium atoms with their very stable electron pairs all very tight around each nucleus have the least capacity for Van der Walls attraction, the positive charge is very effectively shielded, as they say, by the tight skin of electrons. But even they will condense into a liquid when the temperature gets close enough to absolute zero. They have so little energy of motion that the not-quite zero imbalance is sticky enough to match it. Helium is not only the least reactive of the elements, it is the hardest gas to liquefy. But, given ridiculously-low temperatures, helium will liquefy. But them, in extremes, even a helium atom can be forced to give up its electrons. An encounter with an iron atom totally stripped of all its electrons (as could happen in a supernova explosion) will result in an exchange interaction—more a rape, really—dominated by the avidity of the ionized iron to take up electrons. The result is the helium atom is stripped of its electrons which plunge into the inner, empty orbitals of the Fe^{+56} ion. This ion, for that matter, is quite capable of stripping a fluorine atom—this is extreme chemistry; super-valence run amok. Back to the regular world of water moving molecules around.

Lipids take up a structure that minimizes the surface tension of water. A very important class of lipids are those with a highly polar end group attached. One very stable configuration of these is the lipid bilayer. All the long hydrocarbon chains are in the center and all the polar end groups are on the outside interacting readily with water.

These bilayers are very important in isolating compartments in living systems. This is a sophisticated example of simple change—the movement and change in history. It all follows the dictates of the internal wavefunctions and their combinations and collapse.

Pattern matching

Hydrogen bonding is also important in genetics, the complementary matching of base pairs in DNA/RNA. Here the bases do have the capacity to H-bond. But, just like the lipid scenario, when complementary patterns of H-bonding couple with water molecules they also eliminate water molecules and move together. Almost all the basic mechanics of genetics is based on the pairing-preferences of the four "bases" which are linearly strung in their millions and billions as the nucleic acids. (Yes, it is little confusing that linking millions of bases creates an acid, but that's the terminology.) Each base has a pattern of H-bonding-capable patches that complement those on just one of the other bases. Nucleic acids form duplexes—two strands lying side-by-side—when each base on one strand finds its complement opposite it on the other; their H-bond patterns zip together like a mini zipper being closed.

A similar, if much more versatile, movement together underlies much of the work of proteins. For instance, the H-bonding of a protein enzyme and its substrate is such as to eliminate water between them and unite—setting the stage for the substrate to change and, no longer fitting so well, be released.

Again, what is calling the shots is not so much the external form of the system but the patter of internal capacities. It is the patterns that are important in biochemistry and genetics, not so much the molecules on which they are being expressed. The patterns flow from storage in nucleic acids to proteins and back to influencing the patterns being retrieved from the nucleic acids. This can be likened to music which can pattern grooves in vinyl, dots on CD's, radio waves from TV antennas, surges of electrons in amplifiers, movement of loudspeaker membranes and pressure waves impinging on the ear and on as patterns of neuron firing to who knows what in the brain. The external is not of primary significance—though necessary as carrier—it is rather the pattern being passed along. We will return to all this later.

What Are Things Made Of?

So, in brief, the answer that quantum science gives to the question, "What are things made of?" is that they are stuff filling in quantum probability forms.

Quantum science has an equally-brief answer for, "What do things do?" They interact by exchanging and sharing bits of themselves with other things. Just as before, the probability of this sharing is a reflection of a quantum probability form or field.

Forces are resultant things in the new science; forces are a consequence of actually sharing bits of self with another. The classical magnetic force, for example, is a result of the quantum probability of absorbing and emitting virtual photons (phantom bits of light that flit beneath the pixilation of reality and thus do not 'officially' exist.

The simple form to a quantum probability field, the exchange of virtual photons that is magnetism, can be simply seen, just sprinkling iron filings around a magnet.

T-SHIRT SLOGAN

So, if classical science can be epitomized for T-shirts as "all is matter in motion manipulated by forces," then quantum science can be aphorized as "all is matter in external motion manipulated by internal probability fields and forms."

All the sciences would actually like to be 'modern' and manipulate quantum, not classical concepts. Physics, of course, is thoroughly modern. Chemistry with its quantum orbitals is as well. Biology, genetics, evolution, neurology, etc. are decidedly not modern. Biochemistry is currently straddling the fence as the quantum revolution slowly makes its way up the scientific edifice.

It is actually very difficult to switch from the classical way of thinking (probability a result) to the new quantum concepts of causal probability—even Einstein refused to accept the implications of the new physics, in the end, and he helped found it.

So, physics and chemistry now tell us that material objects are made of stuff and probability forms.

Here is another difference between the classical view and the modern: in the new physics, anything that is not forbidden is compulsory, it will happen. Something has either zero probability, or it has a non-zero probability. And even very small probabilities can be significant as they very occasionally get picked.

Let us assume that the same holds for all the other sciences—which are founded on physics and chemistry after all—before returning to our example of applying the concept to protein folding.

Quantum Probability Forms

We have seen the power of quantum probability—holding up aged stars by 1–1=0 alone. Now we are going to look at quantum probability on a more subtle level. For, as we shall see, it is the sophisticated manipulation of quantum probability that underlies—in an internal sense—the marvelous phenomenon of life.

But before we get to living systems, we have to start at the very bottom and work our way up to it as is appropriate in the scientific, bottom-up approach to deconstructing our universe.

The power of quantum probability underlies the less dramatic, but essential, exclusion that gives the elements such as carbon, oxygen and gold their different chemical properties.

For instance, a hydrogen atom is not just an electron and proton near each other. What makes all the difference is the 1s orbital, an intangible quantum probability field with a ball-like form. Quantum mechanics calls this aspect of the hydrogen atom an "internal extension" to distinguish it from the more familiar external extensions in space and time.

The 1s orbital is what gives the hydrogen atom all its character—it is a quantum probability form that is reflected in the overall form to the history of the atomic electron—what the electron does. And chemistry is all about what electrons in atoms do.

All the great difference between the remarkable chemistry that hydrogen atoms participate in—think water—and the null set of helium's relationships is a simple consequence of the fact that hydrogen has a "dissatisfied" singlet electron, while helium has a highly satisfied set of paired electrons. Two electrons in one orbital: one fitting this way, the other fitting that way. And, while the probability of two electrons being in the same state is 0%, the probability of being in the paired state is almost 100%. For a helium atom at room temperature the probability is exactly 100%—helium is totally indifferent to chemical sharing of electrons. Only being totaled in a violent collision can smash the electrons away and this takes a very high temperature, such as in the sun's furnace where even helium is fully ionized.

What are Little Things?

Significantly different from any classical concept is that the total-empty orbitals are just as significant in quantum chemistry as the occupied ones are.

For, even though an empty intangible quantum probability form (QPF) might seem to not belong in considerations of material objects, they are just as much a part of objective reality as the filled ones are. Just ask a chemist if empty orbitals play a role in the behavior of a hydrogen ion or the iron atom at the center of blood-red hemoglobin.

Furthermore, the "size" of an atom (those little colored balls that get tinker-toyed in chemistry) reflects the orbital's sphere-of-influence, not that of electrons and protons. Consider the atom scaled up enormously. The 1s orbital is now the size of a dark and empty Yankee Stadium. The proton has inflated to the size of a baseball at center field. The electron is as a brilliant, but tiny, firefly leaping from spot to spot so much faster-than-the-eye-can-see that the bowl of the stadium is filled with a misty glow, very bright near the baseball but hardly noticeable at the cheapest seats.

If the electron-firefly leaves the stadium, the remaining hydrogen ion is as a dark stadium with a baseball in the middle. But that emptiness is permeated by a quantum probability field, and this is what gives acid its kick.

So classical and quantum physics give different answers to the question: what is a hydrogen atom made of?

The classical answer is: an electron and a proton.

The quantum answer is: ditto, plus a set of quantum probability forms. Some of these QPF are full, some are half-full, and the rest are empty.

This holds for all the elements: they are composed of electrons, nuclei and quantum probability forms. The same holds for molecules in quantum chemistry—which involves a molecular wavefunction—and macromolecules in quantum biochemistry.

Orbitals are perfectly described by complex numbers and, if you have ever seen the Mandelbrot set you have seen the form-making capacity of complex numbers at work.

Providing QPF

We have already rejected the 'lock and key' concept of molecular binding and have embraced the quantum concept of things leaping in and out of quantum probability forms. It would be interesting to know just how close a substrate has to come to its enzyme before it teleports into the highly-probable bound state.

Consider again our hydrogen role-model. One way that we can translate that fearsome-looking quantum equation of the atom is to say that the proton provides a quantum probability form for the electron to fall into. It is an enabler.

The electron, on the other hand, controls the probability of what the nucleus will do. For, when a helium atom collides head on with another atom, it bounces off of it just like a solid billiard ball as in the classical picture.

The quantum view is a little more sophisticated: the attempt by the electrons of the target to enter the filled orbital of the helium atom is repelled with absolute rejection, by the power of the utter impossibility of this ever happening, a power of rejection that is the sole support of elderly stars. No wonder people considered atoms as little tiny bits of impervious solid stuff for such a long time; and did very well with the concept as it is a good approximation in simple circumstances.

Newton's insight still holds—equal and opposite reaction. The helium electrons also recoil in horror at the thought. At room temperature, the probability that the helium nucleus will follow along with these retreating electrons is 100%—the nucleus is constrained by the quantum probabilities provided by the electrons, just as much as the electrons are by the nucleus-provided orbitals.

This, in our example, is as if the baseball conjures up an empty Yankee Stadium; and if a pheromone attracts the fireflea, the whole stadium-baseball follows diligently along. The annals of quantum physics are filled with such odd-to-the classical mind phenomena.

Chemistry is all about providing quantum probability forms for other systems.

SMOOTH AND BUMPY

When free hydrogen atoms meet free oxygen atoms there is nothing to prevent their almost instant embrace. They slide right down the path of least resistance—least free energy in chemical parlance—to bonding as a water molecule.

We can mix hydrogen and oxygen molecules at room temperature, however, and nothing will happen. The gas mixture is quite stable—no water is formed. Even though a water molecule is by far the state of lowest resistance, the path to that state is not a path of least resistance. For the molecules are in a quite contented state. There are no unpaired electrons and all four atoms are in the noble gas configuration. Before the atoms can interact, they have to separate from each other—chemical bonds have to break so they can reform.

The path to this intermediate state is one of very high resistance—very low probability. There is a big bump in the road so the molecules stay intact and the gas mixture is stable. One way over this hill is heat; the hot molecules now have enough kinetic energy to smash each other into atoms. The atoms can now avidly combine. The excess energy is released and heats the gas even more; more smashing and rearranging; more heat released etc.—a runaway chain reaction. Spark a mix of hydrogen and oxygen and you will get an explosion.

In terms of probability, room temperature molecules of oxygen and hydrogen gas have an almost zero probability of making it over the barrier. The situation is just like that of the spontaneous decay transformation of a uranium atom by emitting an alpha particle to a state of much lower free energy. But the path to freedom has a big bump in it. The alpha particle moving through the center of the nucleus interacts with the other nucleons and is strongly attracted to them all. As it is in the center and surrounded, however, the mighty pull in one direction is balanced by an equally mighty tug in the opposite direction. The titanic forces are totally balanced all around and the alpha particle sails on through unimpeded.

At the edge of the nucleus, however, this balance comes to an abrupt halt: the alpha is still being pulled mightily backwards but there is no longer any pulling in the opposite direction. There is a surface tension, similar, if vastly greater, to the force that beads water on wax paper. This is the barrier, the bump in the road to a more stable state. So low is the probability of escape—so tiny is the wavefunction just outside the barrier—that the alpha has to hit the barrier trillions upon trillions of times before it has an appreciable chance that the random operator will smile in beneficent fortune and pick escape for once. While we have drastically simplified the complexities of both atomic and nuclear rearrangements, these general concepts will be sufficient for our purposes.

When the dust settles, the hydrogen and oxygen atoms are in water molecules—they made it through the high-energy intermediate phase riding the crest of the explosion. This is one way over a hump preventing systems from following the path of least resistance. Later we will discuss other, less explosive ways of overcoming such barriers to systems rearranging into states of low resistance. I mention it now only because the straight-downhill interaction of free atoms is somewhat of a rarity in nature; most of the interesting big-picture interactions involve bumps in the path of least resistance.

CATALYSIS BY PROVISION

One of the key differences between living and non-living systems is that, while the wavefunctions involved in the structure of non-living systems are relatively static, living systems are anything but static.

We will start off with the simple concept of systems manipulating other systems by providing wavefunctions—paths of least resistance—for them to follow. The manipulated system is no longer directly dependent on natural law to provide a wavefunction. The system doing the manipulating is the generator.

In both cases, of course, the final step is the same, the collapsed wavefunction—be it natural or provided—has a probability density that will be the actual density given sufficient time and numbers involved.

Nature, of course, has the ability to "do organic chemistry"—molecules get manipulated in their interactions with others. High-energy processes—a spark in the experiment, lightning and solar ultra-violet in the primordial environment—initiated condensations of simple molecules such methane, ammonia and hydrogen and formed a whole mix of organic molecules including simple aminoacids. Today, of course, any products of natural metabolism are quickly swept up by living systems or destroyed by the omnipresent oxygen. But in the pre-biotic world, this would not have been so, and nature-in-the-raw is expected to have populated the early world with a wide variety of simple organic compounds.

It is only relatively recently that chemists have realized just how complex a "metabolism" natural law alone is capable of generating. The pre-biotic history of the earth could have provided many of the components of life—such as simple sugar and aminoacids—along with molecules with the ability to energize transformations such as high-energy pyrophosphates, iron-sulfide compounds, and thioesters. All of which are still to be found at the core of life's current metabolic activity. We can also expect that chemical catalysis was also involved in smoothing the way for these chemical changes to occur.

Catalysis involves providing a wavefunction so reactants can change into products. In the molecular realm, the measure of resistance is called the Gibbs free energy and chemical change follows the path of minimum free energy.

We have already noted that a bump in this path to least resistance occurs when the chemical change involves an intermediate. The block occurs when this intermediary stage has a higher free energy than either reactants or products.

One way around this block is to raise the energy of the reactants—heat or radiant energy are a few of the possible ways. Heat accelerates chemical interactions but is seldom used in living systems. Light, like heat, is capable of energizing many chemical transformations. While visible light energy is used for a lift in photosynthetic systems, this is a sophisticated level of organization.

Only ultraviolet light has much impact on non-living systems and that influence is usually disruptive. Iron atoms, however, can absorb UV and enter a relatively stable excited state—activated ferric ion—that can drive many chemical interactions such as the metabolically-significant high-energy thioesters. These are still to be found at the core of metabolism, and they are thought to have been the first systems that could drive the formation of ATP.

Thioesters breaking up is one of the few chemical transformations whose free energy release is greater than that for ATP breakup—thioester breakup can drive the synthesis of ATP from ADP. Such availability of thioesters provided by activated iron can be expected to play a role in the early proto-metabolism of massive china clay beds fed by both by a black smoker and the surrounding sea water. Most black smokers are in the deep ocean where plates are pulling apart from each other with magma welling up such as all along the mid-Atlantic ridge today.

Such driving of chemical transformations by ATP or thioester breakup is very common in living systems and is well-documented in current science. The energizing system plunges down a path of least resistance and is coupled to pushing the other system up a path of least resistance—making it go in the opposite direction. As noted, paths of least resistance are described by internal natural laws, and internal laws are always reversible (it is the random collapse that is the irreversible step that clicks time ahead).

This coupling of two systems—one going down and the other going up—involves external interaction—there is a physical connection between the two. In this sense, it is an external phenomenon—which is why classical science handles this aspect of living systems very well. It is still a vertical phenomenon in that it can power interactions on many different levels of sophistication. ATP breakdown, for instance, powers all sorts of interactions in the uphill direction on many different levels in the material hierarchy—ions, molecules, macromolecules, spindle construction, cell division, etc.

In classical science, a surface is well-defined as solid boundary. A complex catalytic surface is such a well-defined solid boundary. Unfortunately for this simple view, the new physics says that there is no solid boundary—what we used to think of as the surface of atoms, molecules, clay, etc. is actually tiny electrons teleporting around in vastly larger extended orbitals. The surface is not really a solid boundary at all. For all that, filled orbitals can be roughly equated with classical surfaces. In catalysis, the filled orbitals that participate in providing a path of least resistance for others can be equated with classical catalytic surface that are not well-defined and somewhat fuzzily located. The providing of empty orbitals in catalysis, however, has no classical analog. In classical science, empty cannot be "real." It can, as I hope you remember from our slit set-up where "nothing" stopped projectiles from reaching their targets—and not just photons and electrons, but "solid" atoms as well.

So providing wavefunctions in catalysis has two basic aspects, only one of which has a classical approximation—proving filled-in wavefunction "surfaces" and providing empty wavefunctions with no classical analog.

See the Appendix for a look at the catalytic ability of clay and its role in the origin of life.

HISTORY OF SOPHISTICATION

This is a diagram of our current understanding[1] of life's history in terms of when each level of sophistication was established and set the stage for the emergence of the next level up.

---

1 Christian de Deuve, Blueprint for a Cell, Neil Paterson Publishers, Burlington, NC (1991), p. 59.

Notes: The horizontal endosymbiosis is the internalization of 'bacteria' that became the ancestors of today's mitochondria and chloroplasts. The eubacteria are the familiar ones; the archae-type are a bunch of oddities that live in the most unlikely places like boiling water and volcano vents. The abiotic Earth was molten at first and without life. With cooling and the advent of the oceans, things quickly got going as natural QPF after QPF was sequentially filled in by the 'calcium effect'.

The most difficult step along this history seems, against intuition, to be that of getting the eukaryote pattern fixed as a player on the scene. Some of this difficulty could have involved many "extinction" events such as mar the continuity of the later fossil record. Extinction events, by definition, are those catastrophes—comets and asteroids being the prime suspects—that radically alter

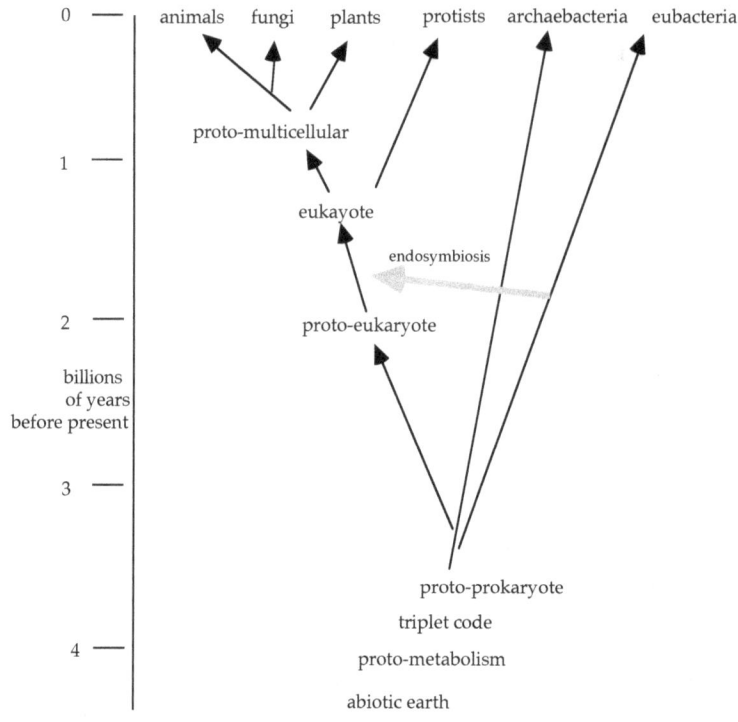

life on earth in relative geological instants. More, of these events in fact, can be expected to have occurred during the first three quarters of history. In general, bacteria are far better at surviving in extreme environments than are higher animals, so we can expect that many promising lineages were extinguished along the way to the one that was established.

"After the appearance of the first endosymbiont-containing protists, evolution once again settled into a relatively static mode, engaging mostly in diversification—endless variations on the same basic themes.... Then some eukaryotes "discovered" the advantages of getting together and pooling efforts. Why it took them so long to make this discovery is not clear. An enhanced interest in sex could be, at least, part, of the answer...."[1] This is an example of horizontal exploration followed by an advance in sophistication. We shall encounter this later in the evolution of the operating systems.

Cambrian explosion

With the maturation of the eukaryote system and the exploration of multicellular possibilities, evolution apparently shifted into high gear in the final quarter of life's history.

The first multicellular organisms appear about 800 million years-before-the-present (MYBP), and, after a period of maturation, the floodgates of innovation opened about 600 MYBP.

In the next 200 million years, all the different phyla of life developed in a tremendous period of development known as the Cambrian explosion.

"About 570 million years ago, virtually all modern phyla of animals made their first appearance in an episode called 'the Cambrian explosion' to honor its geological rapidity. The [fossil record in the] Burgess Shale dates from a time just afterwards and offers our only insight into the true range of diversity generated by this most prolific of all evolutionary events. ... the fossils from this one small quarry in British Columbia exceed, in anatomical diversity, all modern organisms in the world's oceans today. Some fifteen to twenty Burgess creatures cannot be placed into any modern phylum and represent unique forms of life, failed experiments in metazoan design. Within known groups, the Burgess range far exceeds what pre-

---

1 Christian de Deuve, Vital Dust, Basic Books, 1995, p. 168.

vails today. ... The history of life is a tale of winnowing and stabilization of a few surviving anatomies, not a story of steady expansion and progress."[1]

As might be expected, the more complex a system, the more possible varieties might be possible. This expectation is borne out—the mechanisms that emerged at this time capable of managing and duplicating systems at the level of sophistication controlling organs as a unified body opened the floodgates of exploration and innovation. Thus, the great difference in the number of species found in simple (pre-Cambrian) life as compared to the number of species found after the Cambrian explosion. The possibilities of organism structure seem to increase dramatically with sophistication, as seen in this chart:[2].

Kingdom	Features	Number of Species
Monera	prokaryote	4,000
Protista	eukaryote	20,000
Fungi	multinucleate	80,000
Plantae	plant	300,000
Animalia	animal	2,000,000

Compared to the billion it took for mature eukaryotes to emerge from the prokaryotes, the Cambrian period involved radical changes in systems taking place over periods of tens of millions of years—an explosion indeed in terms of speciation.

The rest of life's history has been variations on the themes initiated during that period and, it is only with the advent of the brain capacity that a radically new level of sophistication can be said to have emerged.

Generating QPF

The good news is that I am now abandoning my anthropomorphic way of describing QPF and the impulse to follow the path of least action. The bad news is that I am adopting a new analogy, that of the computer.

Quantum science states that the objective world is stuff filling in quantum probability forms.

A bacterium certainly fulfils this expectation. There are millions of protein-catalysts in the bacteria. Each of the 10,000 varieties is providing a QPF to the overall QPF that is the bacteria. The stuff of the bacteria flows through these QPF, like pipes and pumps in a chemical factory.

The stuff fills in this composite QPF and we see bacteria. While the stuff is constantly changing the bacterial form, reflecting the QPF in the usual way, remains constant.

We are now going to liken bacteria to a quantum-style computer.

First is the fundamental difference between a computer operating system and the programs that run 'on' the operating system.

I am currently running Mac OS X v. 10.3.8 on my PowerBook. My word processor, MS Word, is running on top of this, as well as many, many other different programs.

For bacteria, the operating system is the triplet-code RNA mechanism of protein synthesis. The programs are linear RNA written in triplet code.

In this section, I am going to deal with the programming side of living systems. In the next section, we shall deal with the operating systems of life and their origins.

The triplet code method of protein synthesis has been extensively described elsewhere.

In essence, a digital RNA code is translated into an analogue protein effect contributing a QPF to the composite whole.

---

1 S. J. Gould, "A Web of Tales," Natural History 10/88, p. 16-23

2 Siegfried, K. G., "The Universe and Life" (1987) pp. 261-3.

Digital: Linear RNA code

Translation into linear aminoacids, folding

Analog: Protein providing QPF for stuff to fall into.

This is the basic triplet code and the aminoacid "machine level" processes they call up. Most of the 'desires' of the aminoacids will be satisfied in the folding—which we will get back to by the end of the book—with a few left out as the active site, the QPF being contributed to the composite QPF. The code is "degenerate" in that different codons translate into the same aminoacid; the chart just lists one and the number of degenerates.

Why just 20 aminoacids, why just 20 'machine codes'. This is like asking why our alphabet has 26 letters. The best answer to this perhaps is that they suffice—the dozen or so phonemes of speech can be covered sufficiently well. For there are only a dozen or so elementary chemical reactions important to life's needs.

DESIGNER CODES

As an interesting aside: As this is being written, scientists are beginning to experiment with designer triplet codes—codes that are translated into aminoacids provided by the experimenter that are not found in nature. Two experiments involve bacteria with altered triplet codons that thrive on fluorotryptophan—a deadly metabolic poison to all universal-code users such as bacteria and us.

"One of these two bacteria with the designation "HR15" grew happily on it. Not only did HR15 thrive on fluorotryptophan, it was poisoned by tryptophan. HR15 is not just a picky eater, but an entirely new type of life, [researcher] Ellington says. [Researcher] Wong agrees, 'HR15 does represent a new form of life because the genetic code is the most basic attribute of living systems' he says. He calls the alteration of the genetic code, 'the ultimate test-tube evolution… we are altering the whole organism.' The public has nothing to fear from these artificial organisms, says [researcher] Schultz… Any bacteria that escaped the lab would starve without the researchers feeding them the unusual aminoacids. Bioethicist Caplan dismisses any charges that the researchers are playing God. He says that the scientists are 'playing man' and doing what people do best—creating new things. 'There's nothing wrong, morally, with inventing things,' he adds."[1]

I must say, I do like that delicate correction in the above—not playing God, playing man! Like Father, like Son.

BASIC PROCESS

There is also amplification: one mRNA can be transcribed into many copies of a protein, contributing many QPF to the composite.

In essence, though, we can describe the basic process of life as

A linear program is run on the operating system.

Quantum probability forms are generated.

There are differences of scale, of course. My Mac has one processor running the DOS while a bacterium has hundreds of thousands of ribosome processors all running at the same time.

So, while each ribosome runs just one program at a time, hundreds of thousands of them are all at work simultaneously. Massive processing of relatively few programs. Such massive, coordinated parallel processing is a major goal of computing science but with little success so far.

So, in quantum science, what is the basic description of a bacterium? It is a Mandelbrot Set-like concatenation of millions of quantum probability forms. The collapsed form to this internal aspect is the form we call a bacteria as stuff rapidly pours through the probability gradients.

---

1 Tina Hesman, Science News, vol. 157, June 3, 2000, p. 362.

The numbers involved are roughly 10,000 genes, 100,000 mRNAs, 1,000,000 ribosomes, and 10,000,000,000 proteins. Each protein is contributing a QPF to the composite final probability amplitude that makes a bacterium so probable.

Condensing this description even further with our computer metaphor, we can say that a bacterium is basically ten thousand linear programs running on one million operating systems generating ten billion analog quantum probability forms. Then there is, of course, the stuff flowing through the probability gradients, like electrons in orbitals of molecules.

So where did all those programs come from? We know pretty well how they are passed down and multiply-copied down the ages—some of our housekeeping genes are almost identical to those in use by bacteria, evidence of a common ancestor.

But where do the programs come from in the first place, what was their origin in the first place. "Who wrote the program?" is the first question asked when a Windows virus spreads like the plague—exactly like plague with email playing the role of carrier rats.

This brings us into the thickets and battles-royal of evolution. What are the origins of the one operating system and the ten thousand programs.

Generalized Schrödinger

We are now going to make some drastic generalizations about the nature of the well-formed QPF found in nature. Well-formed, as noted earlier, in that they are relatively long-term and stable forms.

First supposition: The Schrödinger equation that describes atomic orbitals is a member of a much larger class of equations that we will call the Generalized Schrödinger Equation, GSE. All well-formed QPF on any level of sophistication have a form that is accurately described by a Generalized Schrödinger relationship.

We will now dissect the Schrödinger equation, that intimidating hieroglyphic that, believe me, accurately describes the orbitals of the atoms. (Solving it, however, is another question entirely.)

A few points to remember: The proton (for hydrogen) or the nucleus, in the appropriate reference frame, is unmoving. The nucleus, in our time frame, is an unmoving, unchanging generator of QPF orbitals for electrons to fill-in. (See the sections on catalysis and enzymes for more sophisticated generators of QPF for others.

Generators of QPF are always unchanging compared to the transient nature of the stuff filling in the QPF. The Law of Large Numbers insists on it. Otherwise, the probabilities would never get a chance to be expressed by the stuff before the probability changed again. Certain francium atoms, for instance, can never be observed simply because the nucleus flips so quickly to another element that the 90 odd electrons only get to make a few jumps in the francium QPF before the nucleus decays and a new QPF is generated.

In the general scheme of things, note for later that the atomic nucleus plays exactly the same role as catalysts, proteins, and RNA programs running on a real and in a virtual operating system. In this sense, understanding the atomic nucleus from top-down is equivalent to deProgramming it.

Here is the monster we wish to tame.

$$-\frac{d^2\psi}{dx^2} = \frac{2m}{\hbar^2}(E - V(x))\psi$$

The reason it looks so formidable is because it is written in the mathematical equivalent of assembly code. This is what the assembly code, the last step in software before its expression in hardware, for adding two registers together might look like:

10100111010001110010010001010001010010100010000010100

Or in hex shorthand, the command to insert the letter 'I' I just typed into my unsaved Word document in memory might look like:

564FA36EE765BA1000FFFFF22237656.

It is clearly impossible to code any but the simplest of programs in either form.

The top level language running on my Mac OSX, on the other hand, probably looks something like this:

CHECK KEYBOARD, CHECK NETWORK, RUN PROGRAM THREADS, RUN HOUSEKEEPING THREADS, REPEAT.

This is the type of language we will be able to tame Schrödinger with. To give you hope, we will end up with a simple relation such as:

$s = I - qp$

Thou Shalt Not

We are first going to do a simple algebra by dividing both sides by the same thing, the wavefunction, or that funny looking psi.

This of course is only allowable if the wavefunction is never exactly zero, which it never is. For, remarkable as it may seem, the 1s orbital of a hydrogen atom in your body actually has a non-zero value on the Moon. It is an extremely small probability and is essentially zero, but it never gets to exactly zero. It is like the zero limit in calculus.

For while the calculus zero is essentially zero, but it exactly. As any careful calculus textbook will say somewhere, when talking about the limit of one-over-infinity equals zero: "But note that it never becomes exactly zero."

At the very heart of calculus is the assertion that: "Nevertheless, as the difference between exactly zero and our essentially zero can be made as small as desired it can be ignored."

The only real difference is that, while you are allowed to divide by the calculus zero, you are not allowed to divide by exactly zero. Ever. Under any circumstance whatsoever. To do such a thing is to declare yourself a non-mathematician and your theories worthy of ignoring from henceforth.

This is why it is important that the wavefunction never be exactly zero, anywhere. Otherwise, our division would be disallowed.

Luckily, unlike almost everything else we have discussed in physics so far, the wavefunction is not discontinuous, it is not pixilated. It can shrink exponentially and infinitely without ever getting to exactly zero.

Let's get ridiculous for a moment to illustrate this seemingly trivial point. As big numbers are more impressive than the small, first we need to define a really, really truly-enormous number. We start with a big number, the familiar googleplex, 10 to the 10 to the 100, or $10^{10^{100}}$

Call this big number, g. Now raise g to the gth power in a tower of stories g high, $g^{g^{g^{\cdots}}}$. This is a really, really big number; call it G.

Now build another tower of G exponents, this time G stories high. $G^{G^{G^{\cdots}}}$. This is our really, really truly-enormous number; call it G.

Now flip it, calculate $1/G$. This is a really, really truly-infinitesimal number that any well-respected calculus major would be happy to call essentially zero, but would happily divide by it if need to. Call this essentially zero, o.

Now you might think that there would not be much room between o and 0. But you would be wrong. For it is proven that there is an infinity of locations even closer to the true zero. And not just a countable infinity, but an uncountable[1] infinity

---

[1] For an explanation of the infinite difference between a countable and an uncountable infinity, see To Infinity and Beyond, Eli Maor, Birkhasuser, Boston, (1986) pp. 61-63

of points between this essentially zero and exactly zero. Infinitesimally close, believe it or not, still has an infinity of infinity of numbers between it and exactly zero.

All this implies that, while the probability of all your atoms deciding to be on the Moon might be 1/G, it is not exactly zero. This can be considered a challenge to advanced technology: to manipulate and magnify such probability of teleportation before all the oil runs out.

Unlike almost everything else in the universe, the wavefunction is not pixilated, it is absolutely and smoothly continuous creating smooth probability gradients even across Planck pixels of space time. The value associated with that pixel then being the average of the gradient across the pixel. This means that the wavefunction can get arbitrarily close to exactly zero even at the far distant reaches of the universe.

In our stadium illustration of the relative sizes within atoms, quantum mechanics tells us that the firefly spends 99% of its time near the baseball, and 99.999999% of its time in the stadium. Yet it also has a non zero probability of appearing on Mars or Alpha Centuri or Andromeda, just for a Planck or two, before reappearing, unwearied by travel, back in your body on Earth.

Quantum mechanics can explain and justify such oddities with hands tied behind its back.

Dissecting the equation

All that was to justify dividing both sides of Schrödinger by the wavefunction, psi. As the wavefunction is never exactly zero, we are allowed to divide by it, and we get:

$$-\frac{d^2\psi}{dx^2}\frac{1}{\psi} = \frac{2m}{\hbar^2}(E - V(x))$$

We are now going to slice this monster into segments, boil each part down to its essentials, then combine them back together. This will take a while, but it will be worth the effort for our final result is a T-shirt equation that should not intimidate any but the truly math phobic.

$$-\frac{d^2\psi}{dx^2}\frac{1}{\psi}$$

$$\frac{2m}{\hbar^2}$$

$$(E - V(x))$$

THE TWIST TENSOR

We shall start with the real-scary looking expression:

$$-\frac{d^2\psi}{dx^2}\frac{1}{\psi}$$

This is actually not as bad as it looks. For Newton's classical formula, $F = ma$, connecting force to inertial mass and acceleration, can also be expressed in this elegant, but complicated way.:

$F/m = d^2x/dt^2 = dv/dx = a$

In words Newton's declaration is that the force, F, divided by the mass, m, equals <u>either</u> the rate of change of the rate of change in position with time, d^2x/dt^2, <u>or</u> the rate of change of velocity, dv/dt, <u>or</u> the acceleration, a.

So, what Schrödinger is describing on the left is the internal 'acceleration' in the form of the orbital, the rate of change of the rate of change in the form of the QPF wavefunction at any point.

And then this 'acceleration in the form' is divided by the value of the wavefunction at that point. We now have a value for the 'acceleration in QPF' per QPF. This value is then negated, it is rotated by 180° on the complex plane.

We shall call this final value a measure of the 'quantum twist' in the QPF.

For the 1s orbital, this twist about as simple as it gets—perfect, spherical symmetry and with no nodes, even at the nucleus. (A node is where the quantum probability is exactly zero.) In music, the 1s orbital would be called the fundamental waveform that fits and fills the degrees of freedom available.

This simplest, most basic filling-in corresponds to the simplest of programs running on a newly emerged OS.

The twist to the 5f orbital, on the other hand, is complex with multi-nodes, and an accurate description of such a convoluted twist is fiendishly complex in the extreme.

And the twists in a simple QPF such as the molecular wavefunction of a water molecule is intricate to describe.

Luckily, as Einstein discovered to his delight when looking for the simplest way to describe his General Relativity, math has these delightful things called tensors (a sort of sophisticated vector involving matrices). And tensors can describe the twists of even the most convoluted and complex of forms.

So, Einstein used to tensors to great effect to describe the way gravitational mass distorts and twists spacetime. See any good book for more info on this.

Now tensors look deceptively simple: they are just a letter with lots of little subscripts and superscript indices that have to be carefully kept track of in detailed calculations; e.g.: $T^a{}_b{}^{b'}{}_{\cdots}$. We can simplify by letting i stand for all the indices: $T^a{}_b{}^{b'}{}_{\cdots} = T^i{}_i$.

Tensors, and matrices of tensors, etc. are quite capable of handling even the most complicated twists to any QPF. This comes from an excellent introduction to tensors (translated from the Russian!):

"There are quantities of a more complicated structure than [real or complex numbers], called tensors... whose specification requires more than knowledge of a magnitude and a direction."[1]

So, we can now do a radical simplification. We will define the Quantum Twist Tensor or QTT, q, as the tensor array that accurately describes the quantum twist in a QPF. This does a nice job of simplification for us, just like the a in Newton's formula.

$$q = Q^i{}_i = -\frac{\frac{d^2\psi}{dx^2}}{\psi}$$

And, as we have no need to actually calculate the twists in various QPF, that's all we need to know about tensors.

Schrödinger now looks a little less forbidding using the QTT instead of that double-integral mess.

$$q = \frac{2m}{\hbar^2}(E - V(x))$$

We are dealing with two systems that are playing very different roles: the proton the generator, g, providing an orbital for an electron, the filler-in, f, to jump into and 'flesh out' over time by the LoLN. If necessary, we can keep track of what belongs to who with the appropriate indices.

I shall not generate clutter with this. If I did, the q would have a little g index while the m would have an f index.

---

1 Borisenko, A.I. & Tarapov, I>E> Vector and Tensor Analysis, Dover, Ny (1968), p. 1.

PENCHANT AND PASSION

Now for the right-hand side of the genius-monster equation. First, we need another rearrangement, using the simple distributive law of algebra, to get a Planck's Constant, h-bar, inside the bracketed term. Then we also shift the 2 inside the brackets.

The right-hand side of Schrödinger now deconstructs into two fragments that we can treat separately.

$$m/\bar{h} \quad \& \quad 2E/\bar{h} - 2V(x)/\bar{h}$$

The first fragment involves mass—which is Einstein-proved to be equivalent to energy.

The second describes the pendulum-like balance between the maximum potential energy stored in the electromagnetic field, a constant, and the kinetic energy of motion which varies with position.

The kinetic energy is at maximum, the electron is moving very, very fast, at the proton-heart of the H-atom, the tiny baseball at the center of Yankee Stadium. On the other hand, the kinetic energy, the velocity, is essentially zero at the far 'edge' of the atom, the worst seats at the very top.

This is why, in classical terms, the firefly-electron does not just get to sit on the baseball-proton no matter how hard it tries to land. For, when it gets to the very center of the field, it is moving so fast it zooms right past the proton and is way up in the bleachers before it can slow down, turn around, and make another lunge to get to center. As electrons never learn or get bored, they can keep this up forever.

As the kinetic energy is a maximum at the center, the PE–KE expression in Schrödinger will be at a minimum at the very center of the 1s orbital. This is the Principle of Least Action appearing in a simple disguise.

PLANCK AND TIME

The little-h under each term is, as earlier discussed, just the conversion factor into the natural units of nature's pixels. This is Planck's Constant over 2π [π. As this is the ratio of the radius and circumference of a circle, h appears as a length, a radius, in Schrödinger. In other equations, the pixilation factor appears as an enclosure, a circumference, as an uncrowned h.

So "inertial mass over h-bar" is describing the inertia per pixel-radius. Same for other two fragments: the energy terms are per pixel radius.

The Generalized Planck's Constant is then the conversion factor, appropriate to the pixilation and timeframe, involved at any level of QPF under discussion.

Now, as mentioned earlier, action is the basic measure of existence. Its pixilation size is given h, Planck's Constant.

For the most fundamental level of reality, then, we can say that the time-frame for particle existence is of the order Planck seconds. So, this is the appropriate scale to use for the electron.

What about an atom of hydrogen? Is the Planck time still the appropriate conversion factor for atomic existence? I think not.

For an atom of hydrogen can only be said to 'exist' in objective reality over time periods measured in pico-seconds. And, while very, very short period of time, it is as an eon compared to the Planck time.

To make this clear, examine closely this computer-enhanced photo of an electron and a proton—taken with a Planck-time, freeze-frame flash camera—and then answer the question that follows.

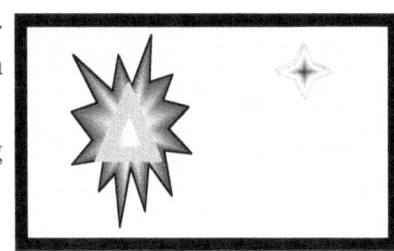

Q. Is this a hydrogen atom? Or is it not a hydrogen atom? And no peeking ahead.

Ready?

A. Yes and No—it's a trick question :-)

The above is actually the supposition of two photos—you can see that the registration is not quite perfect for the two electrons at upper right.

One photo is of a hydrogen atom bonding together a G to a C nucleotide in a 50,000,000-year old sample of dinosaur DNA. This is clearly an atom that can be said to exist.

The other photo is of the inside of a TV tube displaying the "Golden Girls." The electron, at 500,000 mph, is moving horizontally from the cathode towards a red pixel on Blanche's heaving bosom. The proton is a tertiary cosmic ray moving vertically downwards at 10% the speed of light. In a picosecond, they will be miles apart. This second photo is clearly not of a situation that can be said to be an atom that exist.

The point is, at Planck time resolution, the two situations are indistinguishable. The concept 'atom' has no meaning over time pixels commensurate with Planck. A movie of either situation, taken at one frame a Planck, and screened at 100 frames a second, would take millions of years before there was any discernible movement in either electron or proton.

So, the conversion factor in the generalized Schrödinger will be different at different levels. As we are not going to actually do any calculations, the technical question as to exactly what values these factors can be left for now.

As a general rule, however, the timescale must allow for a QPF to be filled in a few score times. With this in mind, we can suggest some approximate time frames for systems with a lots of QPF that have to be substantially filled-in.

PARTICLES: Planck-secs	**ATOMS**: pico-secs
RIBOSOMES: mille-secs	**CELLS**: seconds
ORGANS: minutes	**BODIES**: hours
FAMILIES: days	**NATIONS**: years
SPECIES: decades	**GENERA**: centuries

Planck's Constant is not just about timeframes, it is a product of time and inertial mass. The inertial mass of a system is a measure of the system's reluctance to alter its current state of motion. We will discuss such reluctance-to-change, in general terms, in the following section.

So, the conversion factor in a generalized Schrödinger will also take into account the scale of the resistance-to-change.

That is all we need to note about the fact that the fragments of Schrödinger we are currently considering involve the 'radius' of a level-appropriate time and reluctance, conversion-to-pixels factor.

RELUCTANCE TO PASSION

First, the simplest of the Schrödinger fragments (using ℏ for h-bar as Word prefers it):

$m/\hbar$

The inertial mass, m, measured in appropriate units, of the electron is a measure of its reluctance to alter its state of motion when tugged at by classical forces (actually moving in quantum probability gradients, of course). We can think of this simply as the tendency of the filling-in system to keep doing its own thing.

In the useful classical terms of the "movement" of the electron in the electric field, this inertial mass is a measure of the tendency of the electron <u>not</u> to respond to the electric force. So, the mass of an electron can be thought of as its reluctance, or resistance, to moving so as to fill-in the QPF orbital.

All systems can be expected to put up some resistance to filling-in a QPF, and this can be called the generalized inertia of that subsystem.

We shall define the quantum inertial reluctance of a subsystem, r, to be this pixilated, generalized inertia. This is the measure reluctance to move as the QPF dictates, not as the free system would if left alone. So, for an electron moving in the QPF of a proton:

r = m/ℏ

We now invert this and, as the inverse of reluctance is passion, perhaps penchant, we now define the Quantum Pixilated Passion, the QPP of the electron, p, as:

p = 1/r = ℏ/m.

Substituting this into Schrödinger, then multiplying both sides by p, we end up with the much simpler-looking equation. Is it not great how math can hide a lot of detail with a few simple symbols!

qp = 2E/ℏ − 2V(x)/ℏ

Quantum Intensity & Satisfaction

We can take this process of well-defined generalizing even further and simplify E−V(x).

This expression is notoriously difficult to solve explicitly. Luckily, we do not wish to calculate it, just understand what the expression is telling us about the way the internal world of the QPF is ordered.

PENDULUM AND PHASE SPACE

All we will need for this discussion is the simple pendulum, a favorite gadget in the elementary physics lab.

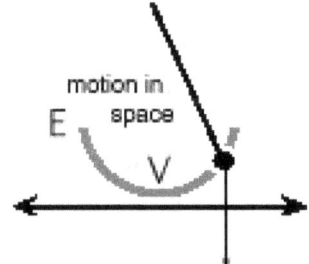

At point E the bob is momentarily at rest, it is not moving. All the energy of the interaction between bob and earth, via graviton exchange, is in the intensity of the interaction. A physicist would say that all the energy of interaction is in the field at this point. This we define as a measure of the intensity of the interaction, I. At point E, all the energy is in I, the potential energy of the interaction.

At point V, all the energy is now in the velocity of the bob. All the energy of interaction is now in the velocity of the bob. No energy is in the potential field, it is all in the velocity of the bob. Note that at V, the velocity is horizontal while the force of gravity is at right angles to it. At point V, and at V alone, the bob's velocity is not influenced by the intensity of the interaction. The bob is in totally-free movement and unencumbered by force. This we define as a measure of the bob's satisfaction during the interaction, s. At point V, all the energy is in s, the kinetic energy of the bob in free and full motion.

At the point where the bob is at in the diagram, at x away from V, some of the energy will be in the intensity of the field exchange particles and the balance will be the kinetic energy of the bob's motion (with horizontal and vertical components) of its attempt to attain complete satisfaction at V again. The bob 'wants' to stay at V, but it is moving way to fast to stay there. Life's like that.

This balance is described an expression that is familiar from Schrödinger. E-V(x)

A better way to describe this back and fore motion (which involves sines and cosines) is as circular motion in a phase space.

The two axes are potential energy and kinetic energy. The movement of the back-and-fore pendulum at variable speed is now a point in this phase space moving at a constant speed in a perfect circle (assuming no friction, which is common). Even at this simple level, it is clear that constant motion of a point in a circle is easier to deal with mathematically than variable side-to side-motion of the actual bob.

Even though a pendulum bob is composed of quintillions of particles each with its own phase space, they all combine into the simple two-dimensional phase space that is sufficient to describe the behavior of the pendulum.

We can now apply this concept to the electron in the 1s orbital ground state, isolated H-atom.

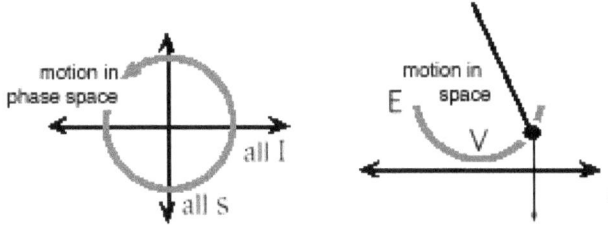

When the electron is at the very edge of the atom, its velocity is zero and all the energy is in the intense electromagnetic field as the potential energy. This value appears in the E of Schrödinger, a constant. This is I, the intensity of the electromagnetic interaction.

At the nucleus, the reverse is true. All the energy is now kinetic and in the motion of the electron, E=V at this point. The electron wants to stay there at the center, but when it gets there it is moving way too fast to stay there. This is the maximum satisfaction of the relationship and at this point, when s=I.

Next, draw a line connecting every point on the surface of the 1s orbital through the center to the point on the opposite edge. There will be a continuous infinity of such lines. Now consider the electron to be moving back-and-fore along every one of these lines at the same time (this only sounds impossible because we are using classical concepts (in a quite valid way) in the description).

As the 1s orbital has circular symmetry with no nodes except at infinity, its motion will be a perfect circle in an infinite-dimension phase space (a common creature in physics calculations). The other orbitals are, like Ptolemy's heavens, combinations of many such circular motions, or epicycles, in phase space. In phase space, the energy oscillations are a multidimensional circular motion.

The maximum potential energy for this situation, a constant, is the E in Schrödinger.

The amount of this energy in the kinetic form, that varies with position from the center, is the V(x) in Schrödinger.

In the helium atom, all we have to do is consider two electrons at each end of the lines through the center. They oscillate happily together. When one is at the end of a line, the other is at the other end. When both are at the center, all the energy is in their mutual motion past each other. Whoosh. Maximum, internally-amplified, satisfaction. Over and over again.

Substituting our more general symbols for intensity and cyclical satisfaction, we have:

$E - V(x) = I - s$

Simple Schrödinger

Putting this back into our deconstructed Schrödinger, we end with the simple, general form that applies to any and all QPF:

$q = (I - s) r$
$= (I - s) / p$

In words, the twist tensor in any interaction equals the reluctance times intensity minus satisfaction, or equivalently, the intensity minus satisfaction over the passion.

This sounds like philosophy and theology, but it is not. Each term has been precisely defined mathematically.

When applied to the hydrogen atom, this general equation gives us back the highly-specialized original:

$$-\frac{d^2\psi}{dx^2} = \frac{2m}{\hbar^2}(E - V(x))\psi$$

Now lets do a little algebra of the internal realm and rearrange the generalized Schrödinger equation to different forms:

$q = r (I - s)$
$r = q / (I - s)$
$I = pq + s$
$s = I - pq$

If you put these into words, you will find a lot of wisdom. What is maximum satisfaction when two fill in a QPF together (a zero is when I = pq)? What is a minus twist tensor (when s is greater than I and r is non-zero). What situations generate maximum reluctance (inertial mass) and what is negative reluctance (when s is greater that I)? etc.

Program Evolution

In the history of Earth, we find that species have distinct beginnings and endings. As Niles Eldridge puts it:

"If the fossil record has anything at all to tell us about the history of life, it is that species of 600 million years ago, or 400 million, 200 and 100 million years ago, are all different from the ones we have on earth today ..... we must further conclude that species undergo a 'birthing' process as well as a 'death'—or extinction—process."[1]

"Nothing in biology makes sense except in the light of evolution,"[2] a sentiment echoed by most biologists.

This 'keystone' element in the edifice of biological thought has gone through its own evolution and development resulting in what is known as the "Modern Synthesis," the 'received view' of contemporary scientists.

Although many natural philosophers such as the early Greek thinkers and Jean Baptiste Lamarck (1744-1829) had toyed with the idea of evolution—the idea that all life descended from a common ancestor—it is Charles Darwin (1809-1882) who, along with Alfred Wallace (1823-1913), is credited with responsibility for the foundations of our contemporary understanding of the evolutionary process. Darwin developed a comprehensive theory that he presented in "On the Origin of Species by Means of Natural Selection" published in 1859.

The theory he presented coalesced many of the disparate facts already catalogued by the exploratory science of that day. "Darwin was able to weave together an interlocking set of hypotheses explaining resemblance among organisms, their patterns of distribution, and their fossil records—a set of hypotheses making sense and providing coherence to a wide body of observation and experience that had accumulated by the mid-nineteenth century."[3]

Darwin's seminal work is the foundation of the modern synthesis. Other elements of the edifice were contributed by Mendel and other workers in genetics and, relatively recently, the explosive expansion of our comprehension of molecular biochemistry.

The disparate elements were first combined to create the modern synthesis by Theodore Dobzhansky in his "Genetics and the Origin of Species" published in 1937.[4]

This is a succinct description of the modern synthesis from a textbook on evolution: "New species usually arise through the accumulation of different genes within reproductively isolated populations of some parent species. These populations become so different that they cannot breed back to the parental population and thus can be recognized as distinct species."[5]

The contemporary view has answered many of the questions about evolution—it has many accomplishments to its credit. There are also, however, some questions that are not answered satisfactorily in the modern synthesis. There continue to be many challenges to the received view.

Well within the scientific mainstream are challenges that arise from recent developments in molecular and population genetics and in paleontology, the study of fossils. A recent review of "The Evolution of Darwinism" described some of these challenges: "One is a proposal that a kind of molecular determinism, rather than pure chance, impels the development of

---

1 Eldredge, N. (1985) Time Frames, Simon & Schuster, p.104.

2 Dobzhansky T., Ayala F., Stebbins G. L. and Valentine J. (1977) "Evolution" (pub: W. H. Freeman, San Francisco), 1977 p.5.

3 Ayala and Valentine, 1979, p.7.

4 Dobzhansky, T. (1937) "Genetics and the Origin of Species" (1st Ed.); 2d Ed., 1941; 3d Ed., 1951: Columbia University Press.

5 Stebbins, L. and Ayala, F. J. (1985) "The Evolution of Darwinism" Scientific American Vol. 253 #1, July 1985, p.13.

variation in DNA. The other is a contrasting claim, known as the neutral theory, that chance governs not only the initial appearance of genetic variations but also their subsequent establishment in a population. A different kind of challenge, based on new interpretations of the fossil record ... known as punctuated equilibrium ... holds that evolution proceeds not at a steady pace but irregularly, in fits and starts."[1]

In the 1940s, Ernest Mayr proposed that transspecific development might occur at a different tempo than subspecific development,[2] a proposal that was developed into the "punctuated equilibrium" theory of Gould and Eldredge.[3] This theory proposed that speciation occurred in small populations and very (geologically speaking) rapidly, an idea that has received a great deal of empirical support.[4]

One of the key concepts in the modern synthesis is that "evolutionary change must be dominantly continuous and descendants must be linked to ancestors by a long chain of smoothly intermediate phenotypes."[5] This idea was challenged by the extreme saltationist view that development proceeded by large jumps through the appearance of fortunate macromutation, the "hopeful monster."[6]

Although this idea was not well received by the scientific community at that time, recently it has:

"Been reborn as a product of the transposition of small regulatory elements of DNA, or by the translocation of large chunks of genome, leading in either case to major changes in gene expression by means of which, according to a flight of fantasy indulged by W. Doolittle, a toad might evolve into a princess with a minimum of intervening millennia."[7] The evidence for this 'quantum speciation'[8], its possible mechanisms[9] and saltationist models of evolutionary processes[10] are now the subject of debate in the scientific literature.

Saltationist views have gained ground in the scientific community promoting the comments: "Quantum speciation of any sort was rejected ... in retrospect, it seems that Goldschmidt deserves posthumous accolades for his steps in the right direction."[11] And, "Quantum speciation entails no major elements not recognized within the Modern Synthesis of evolution. The new view simply differs in its emphasis on particular elements and in its implications for large-scale evolution."[12]

---

1 Stebbins, L. and Ayala, F. J. (1985) "The Evolution of Darwinism" Scientific American Vol. 253 #1, July 1985, p. 72.

2 Reviewed: Mayr, E. (1982) "Speciation and Macroevolution" Evolution 36(6) p.1119-1132.

3 Gould S. J., and Eldredge N. (1977), "Punctuated Equilibria: The Tempo And Mode Of Evolution Reconsidered" Paleobiology 3 pp.115-151

4 Stanley S. (1982) "Macroevolution and the fossil record" Evolution 36(3) p.460-473

5 Gould S. J. (1982) "Darwinism And The Expansion Of Evolutionary Theory" Science216 p.380-387

6 Goldschmidt R. (1940) "The Material Basis of Evolution," Yale University Press

7 Robertson M. (1981) "Gene Families, Hopeful Monsters And The Selfish Genetics Of DNA" Nature 293 p.333-334

8 Schopf T. (1982) "A Critical Assessment Of Punctuated Equilibria: 1. Duration Of Taxa" Evolution 36(6) p.1144-1157

9 Rose M. and Doolittle W. (1983) "Molecular Biological Mechanisms Of Speciation" Science220 p.157-162

10 Doskocil J. (1983) "A Model Of Stepwise Evolution Of Higher Eukaryotes" Folia Biologica (Praha) 29 141-155

11 Stanley S. (1981) "The New Evolutionary Timetable" Basic Books, NY, p. 135

12 Stanley S. (1981) "The New Evolutionary Timetable" Basic Books, NY, p. 166

There are those who have pointed out that Darwin himself can be considered a "punctuationist"[1] and, naturally enough, there is also a spirited defense of the Modern Synthesis or neo-Darwinian thought as is.[2]

The modern synthetic view predicts that, if the fossil record were exact enough, as the paleontologists dug and sifted through geological time, they would see a gradual drift, the transformation of a species into another.

Niles Eldridge, curator at the American Museum of Natural History and one of the developers of the theory of punctuated equilibrium, recalls his first experience of the difference between his expectations based on the modern synthesis and what he actually found in his explorations of the fossil record of the trilobite Phacops rana:

"But that's not what's there ... in the entire 8 million years ... the greatest (though not the sole) amount of modification wrought by evolution in the Phacops rana stock was the net reduction from 18 to 15 columns of lenses. Hardly prodigious, this degree of anatomical retooling falls well within the normal bounds of 'micro-evolution'... We see something out of whack with prevailing expectations... as we climb up those rocks and check those samples, over what must be, in sum total, a 3-or-4 million year period, we see some oscillation, some variation, back and forth ... but no real net change at all ... This is the first element: simple lack of change. Stability, or stasis as [Stephen Jay] Gould and I began to call it. And the second element in this pattern is the apparent suddenness of the change: when it does come, evolutionary modification seems to be abrupt, an all-or-nothing sort of affair."[3]

The quantum nature of the fossil record had been quite apparent from the very beginnings of modern paleontology.[4] Darwin asserted, and this view has been incorporated into the modern synthesis, that this was an artifact. That, because the fossil record is incomplete, we gain the impression that a quantum change has occurred when in fact, if a temporally-complete selection of remains had been preserved we could then see the actual, gradual transformation occurring.

Darwin was very clear on this point:

"I have attempted to show that the geologic record is extremely imperfect; [a long list of reasons why]. All these causes taken conjointly, must have tended to make the geological record extremely imperfect, and will explain, to a large extent, why we do not find interminable varieties connecting together all the extinct and existing forms of life by the finest graduated steps. He who rejects these views on the nature of the geological record will rightly reject my whole theory."[5]

This is why there is almost a sense of relief in the paleontology world that the record is not complete. In the 1950s, one book on evolution clearly expressed it: Thank goodness, the fossil record is not complete![6] Why would such a strange sentiment exist—gratitude that the experimental data was incomplete? The simple reason is that, as already noted, for science to progress there has to be the ability to order and classify the complexity of nature. If the differences between individuals cre-

---

1 Rhodes, F. (1983) "Gradualism, Punctuated Equilibrium and the 'Origin of Species'" Nature 305 p.269-272

2 Charlesworth B., Lande R. and Slatkin M. (1982), "A Neo-darwinian Commentary On Macroevolution," Evolution 36(3) pp. 474-498

3 Eldredge, N. (1985) Time Frames, Simon & Schuster, pp. 69-70. Later in the book (p. 82-83), the author describes later work which developed his view that this abrupt change in the trilobites occurred over a period of thousands of years in another area with a geographic change accounting for the apparent abruptness.

4 The use of the quantum concept is not the same way as used by George G. Simpson (Simpson 1944) where it refers to rapid, rather than sudden, evolutionary change.

5 Darwin C. (1859) "On the Origin of Species by Means of Natural Selection or the Preservation of Favored Races in the Struggle for Life" New American Library, reprinted 1958, first ed. p. 341.

6 Cain, 1954, as noted in Niles Eldredge's "Time Frames," p. 69.

ated a continuum it would be impossible to consign individuals into larger groupings, the neatly ordered, set of inter-nested boxes labeled with Latin binomials, into which we have been able sort individual organisms since the time of Linnaeus.[1]

As Darwin noted, the fossil record most definitely did not show the gradual transformation of one species into another:

"Why then is not every geological formation and stratum full of such intermediate links? Geology surely does not reveal any such fine graduated organic chain; and this, perhaps, is the most obvious and gravest objection which can be urged against my theory. The explanation lies, as I believe, in the extreme imperfection of the fossil record."[2]

There have always been, however, since the very start of the debate, certain paleontologists who have disagreed with this 'incomplete' interpretation of the quantum nature of the fossil record. They are known collectively as 'saltationists.'[3] Although in many ways their ideas can be very different, basically, they all maintain that evolution proceeds by leaps, sudden jumps from one state to another. This classification of scientists is broad, encompassing the early eighteenth century catastrophism of Georges Cuvier and the 'hopeful monsters' of geneticist Richard Goldschmidt of the 1940s.

dePrograming

The differences between species of bacteria are not to be found in the operating system for they are all identical (we use exactly the same OS ourselves in our organelles). It is the programs running on the OS that are different.

Evolution, then, must involve the evolution of programs—the evolution of programs that generate quantum probability forms when run on the OS. The task of the scientist, then, is to deconstruct the programming of living systems, or as I like to call it, dePrograming nature.

So, the struggle for existence happens first on the internal programming level, second in the external world. Moreover, the criteria for success in the internal world are quite different from those required for success in the external world.

Classical pictures of evolution have focused, of course, on the external stuff. What the material, the atoms, the molecules are doing. It has no concepts that can deal with why this flow of stuff is so probable that it happens all the time.

The inherent improbability, according to classical science, of many of the processes known to have occurred in evolutionary history has troubled many workers in the field.

One provocative book, in its attempt to solve this problem, in 1981 created a tremendous stir in scientific circles. In Britain, the reviews of the book A New Science of Life: The Hypothesis of Formative Causation covered both extremes:

Nature, the preeminent international science journal declared it "the best candidate for burning there has been for many years" while the New Scientist, a feisty news magazine, stated, "It is quite clear that one is dealing here with an important scientific inquiry into the nature of biological and physical reality."

The reason why the book created such a stir was that, after listing in "Some Unsolved Problems of Biology" the inability of classical theory to deal with questions of morphogenesis, evolution and behavior, Dr. Sheldrake introduces a new causal factor into the scientific picture of how the world works. He postulated a morphogenetic field: a non-energetic tem-

---

1 Such thinking is not only relevant to the fossil record, but also in the contemporary world. If the processes of microevolution—the development of variation within a species—are the same as the processes of macroevolution—the development of species—we might reasonably expect to find a continuum of variation in the modern world. It could be argued that natural forces have worked to separate populations into the temporally real, if historically transient, 'reproductively coherent communities' we call species. But it does not seem unreasonable to expect to find areas of life where such separation of the continuum of variation is still occurring and the continuum still exists. Such is not the case, however. Clear-cut species boundaries are the rule.

2 Darwin C. (1859) "On the Origin of Species by Means of Natural Selection or the Preservation of Favored Races in the Struggle for Life" New American Library, reprinted 1958 p. 280

3 From the Latin saltareto leap

plate or blueprint that guides physical, chemical and biological systems so that only one result occurs out of the many that are equally possible energetically. He uses the following analogy:

"In order to construct a house, bricks and other building materials are necessary; so are the builders who put the materials into place; and so is the architectural plan which determines the form of the house. The same builders doing the same total amount of work using the same quantity of building materials could produce a house of different form based on a different plan. Thus, the plan can be regarded as a cause of the specific form of the house, although of course it is not the only cause: it could never be realized without the building materials and the activity of the builders. Similarly, a specific morphogenetic field is a cause of the specific form taken up by a system, although it cannot act without suitable 'building blocks' and without the energy necessary to move them into place."[1]

It would seem that the 'morphogenetic field' proposed by Dr. Sheldrake is already a part of modern science—it is equivalent to the composite quantum probability form.

External Evolution

In classically-based evolution, evolution is left up to 'chance and accident.' Yet bacterial life appeared on the cooling earth not long after the oceans were finally stable and established.

We have already gone to great pains to show that such classical concepts of 'probability' have been totally repudiated in physics and chemistry.

Classical evolution then is definitely assailed from below by the quantum probability revolution in physics. It is also under attack from above, for classical concepts lead us to expect life to be highly, highly improbable. In classical science, a construct as sophisticated as a living cell is highly unlikely. As Fred Hoyle put it, the emergence of living systems is about as likely in classical physics as a hurricane sweeping through a junk yard assembling a fully-functional Jumbo Jet from the bits and pieces scattered around there.

Just as the chance of junk colliding in just the right way to form a fuselage is small, so, in classical physics, is the chance that atoms and molecules will congregate in just the right way to form cells, organelles, tissues, etc.

This view of evolution permeates all of biology, all of genetics, all of the brain sciences.

"Nothing in biology makes sense except in the light of evolution,"[2] is a sentiment embraced by most biologists. The classical science system put such great emphasis on fighting off a teleological explanation of evolution—that there is a purpose and a plan behind the origin of species—that biologists have gone to the opposite extreme and adopted the concept that there is no underlying organizing factor to evolution.

All scientists believe, of course, that the phenomena of Nature can be understood and that there are still many things that our science has yet to figure out—they would be foolish indeed to do "research" if they didn't believe there was anything left to discover.

So, it is strange that, while no one is saying that electrons and protons behave in a totally random chance-and-accident manner in the formation of simple atoms, many biologists are stating this to be the case for the much more complex rearrangement of the genetic molecules that occurs in the historical development of species, genera, etc., during evolution.

---

1 Rupert Sheldrake, A New Science of Life: The Hypothesis of Formative Causation, J. P. Tarcher Inc. Los Angeles, distributed by Houghton Mifflin Co., Boston (1981), p. 71.

2 Dobzhansky T., Ayala F., Stebbins G. L. and Valentine J. (1977) "Evolution" (pub: W. H. Freeman, San Francisco), 1977, p.5.

"Biologists think it essential to avoid asserting anything vitalistic. The only way to do this is to deny any vestige of entailment in evolutionary processes at all. By doing so we turn evolution, and hence biology, into a collection of pure historical chronicles, like the tables of random numbers, or stock exchange quotations."[1]

The reality of evolution, of course, is no longer a point of debate. There is such clear and abundant evidence that all life is lineally connected that it can be accepted as an established fact. Life is lineage; we are all connected through our ancestors.

Looking back some million years ago our lineages merge with those of the great apes, further back, with the primates, the mammals, the reptiles etc. This vast, interconnected lineage took its time developing: a few hundred million years or so after the molten Earth cooled off for basic bacteria-like organisms to develop from simple chemicals; another billion years or so for the development of complex cells; another billion for multicellular plants and animals to emerge; just a few tens-of-millions more for all the current diversity of living systems to be established; and the last half billion or so for the emergence of creatures aware enough to wonder about how it all happened.

The chance of even one specific protein being formed out of free aminoacids is of the order 1 in 10^{300}, while the odds of proteins etc. coming together to randomly form a simple bacterium are on the order of order 1 in $10^{34,000,000}$. Events with such odds against them could never be expected to happen in our universe that is only 10^{17} seconds old.

Whatever events occurred during evolutionary history, it is clear that such odds were never encountered at any step on the road from bacteria to man—certainly not a series of such highly improbable events. Rather:

"One can assume that life arose through an enormous number of small steps, almost each of which, given the conditions of the time, had a very high probability of happening ... a multiple-step process that relies on one improbable event's following another is sure to abort sooner or later."[2]

This quote is from Vital Dust, the best book I have come across that conveys the Big Picture. It covers everything known about what actually happened, from molten earth to human culture.

The perspective is classically-based, however, so cannot deal with its own conclusion that each of the many steps along the way "had a very high probability of happening."

The inclusion of quantum probability in the conceptual armory allows us to approach this central question. Classical biology is as incapable of dealing with why life "had a very high probability of happening" as classical physics found itself incapable of dealing with the two open windows as bulletproof as steel in our execution illustration of the slit experiment.

Just in case the above makes a theologian smirk of satisfaction in this swing in favor of God!, we should note that the new physics is just as inimical to one of many a religion's favorite axioms: God is in Control and knows what is going to happen.

For the new science asserts that it is impossible for even God to know which slit an electron will choose to go through. He can know the probability to the nth decimal place, but He cannot, in principle, know which slit will be picked by the autonomous electron. While God might manipulate probabilities in history like a divine psychohistorian, God is not going to know exactly what humans are going to choose to do. We have creative freedom and generate our own probabilities.

So, if there is blame to be laid for the historical misery of humanity's history it is either that God failed to make the probabilities of success great enough or that humans made very unwise, highly unlikely choices. "God is in Control" of what happens is totally incompatible with the quantum view of the world and must be ejected along with the other classical concepts we discarded earlier.

---

[1] Robert Rosen, Life Itself, Columbia University Press, NY (1991), p. 256.

[2] Christian de Duve, "Prelude to a Cell," The Sciences, Nov./Dec. (1990), p. 24.

Both science and religion aim to describe the 'truth' about the reality we jointly inhabit. So eventually, if both get better at the task, they are going to end up converging. Right now, however, they are often so far apart, the concepts so black-and-white, so utterly contradictory, that it makes sense, in all current cultures, to ask acquaintances, "Do you believe in science? Or religion?"

I hope, I assume, that in some culture to come, this question will be as silly as asking, "Do you believe in physics? Or chemistry?"

Making it as black and white as it gets: religion insists that evolution is determined by fiat; classical science says that it's all chance and accident. The new physics suggests they are both wrong: natural law determines the probability of things happening; the rest is up to time and things filling in the probabilities.

Back to bacteria who, to misquote, God must love dearly because there are so darn many of them to make us look like afterthoughts.

Internal Survival

We shall leave the origin of the One Basic Operating System of life, the triplet code method of protein syntheses, until the next chapter. Here we shall focus on the origin of the evolution of the programs.

What do programmers do when they write new programs? Really efficient ones reuse the same code over, tweaked for different purposes.

So, the origin of new programs in bacteria can be expected to involve mixing, matching, and a lot of duplicated code with slight differences.

Both processes are well-documented in bacteria. Proteins fall into distinct lineages with ancestral connections. They also mix their DNA stored programs with other bacteria in a simple form of sex. So, we can envision new programs as starting with new combinations of subprograms already in use.

But new raw code is not sufficient, it has to pass certain internal criteria.

For a program to do well in the environment provided by the operating system it will have to follow simple rules. We can illustrate this with my recollections of programming with MS Basic on the Mac XL.

The correct grammar must be followed for a program to run:

```
10  GOTO 20    10 GOTP 20
```

The correct syntax must be followed for a program to run:

```
10 GOTO 20    10 GOTO GOTO
```

There must be a start—easy, and there must be an end—tricky, as endless loops are all too easy:

```
10 GOTO 20    10 GOTO 10
```

There must be nothing that crashes the OS:

```
10 DIV 1 BY 1+1    10 DIV 1 BY 1-1
```

It must be elegant and use code efficiently. This is a higher level of programming success. You do not recode a wheel each time you need one; you do it once, then you call it up as a subprogram. When I type on my Mac, something like the following program runs:

```
CALL keystroke detected
 CALL letter typed
  CALL send ASCII to screen RAM
   WRITE pixel pattern to screen
CALL keystroke detected
```

When a subprogram gets called a lot by 'higher' programs, it becomes relatively unchangeable and fixed down the generations. Most of our housekeeping genes, for instance, haven't changed much in a billion generations since the mud days. Changing them would be like changing the ASCII code for 'e'—impossible.

Only at the very top levels is program experimentation allowable.

We have already noted the fundamental principle that, in quantum science, what things are and what things do is all matter in motion in quantum probability forms.

This holds for bacteria. The myriads of proteins in a typical bacterium are each contributing a QPF to the mix. The overall composite of these, when filled in, is what one could call a healthy bacteria.

Metabolism is not a static thing; rather, foodstuff, etc. flows through the composite bacterial QPF. Each metabolic step can be likened to pipes in a chemical factory leading from reservoirs to reactors to another pipe; the width of a particular pipe reflecting just how many of that particular QPF are in the composite QPF. The ATP reservoir is small, for example, but it has a huge inflow pipe and thousands of small output pipes.

This is basic metabolism. But there is another level of control adjusting the size of the pipes and thus regulating the chemical transformations. This involves the regulation of transcription of housekeeping genes. Then there is a level that regulates this.

At the top is a program running that we can call 'life is good.' This Program is running when bacteria have food and are growing and dividing.

There is one more level of programming control. If the food disappears, the bacterium flips its state and becomes a spore, tough and resistant and awaiting better days.

The bacterium has allowed itself to be programmed by its environment—the environment is the final programmer of a successful bacteria. Survival of the fittest can be rephrased as survival of programs capable of being programmed by the environment.

The best analogy to all this is my computer. It sits there running dozens of threads doing who-knows-what until I hit a key. This simple input causes a cascade of changes to the RAM, to the video memory, to the pixel patterns on the screen.

At the bottom is the basic code that actually runs the machine. This is almost impossible to write a program with. So higher languages, such as C++ and on up, are used. At the top are the main programs, such as the MS Word virtual environment I am writing in. This sits right on top (for most of the time, as discussed later in Sleep).

In one of these 'higher' languages, we can say our little bacteria is running a simple program right at the top of the hierarchy:

WHILE input **IS** good

   RUN life is good

ELSE RUN batten down the hatches

When famine strikes, in a very short time the composite Quantum Probability Form that is the thriving bacterium that is being generated by the "life is good" program switches to a composite form that is the spore. The stuff automatically falls into the new probability gradients and a spore results. Just three lines of code are sufficient for the trick.

The environment and such a bacterium are in a relationship just like my Mac and me. To the programs running in the Mac, I stand in the position of User. The Mac housekeeping programs are all busily running, layer upon layer, busily shifting stuff around, until I hit a key, tap, and Notice Must Be Taken and the book progresses.

The bacterial programs are the same, happily humming along until the environment goes, tap, and Notice Must Be Taken, and the BECOME A SPORE program starts to run. This state continues until a, tap, from the environment, and the BECOME A NOT-SPORE program starts to run. The environment is in the role of User to the bacterial program.

This motif holds throughout genetics. When, during development, a cell differentiates into a liver cell, say, it is because it received a tap from an organ program, which, to the cell program, stands in the position of User. To the organ program, of course, the liver cell program is just another trusty subprogram it makes many calls to.

SYNTAX CHECKER

Modern bacteria are so sleek, their programs so optimized and elegant, that it is not implausible that the final perfection of the bacterial form involved an efficient way of weeding out programs before they ever had the chance to run on a real operating system.

This is like a virtual reality. In this virtual reality, the novel programs could be virtually run with virtual consequences. Only programs that do not commit a faux pas are released to the real world and to get run on a real operating system such as a ribosome.

Lots of novel programs and combinations of them with the old can be released into the virtual OS and run. Only those that are deemed "well-formed" are allowed to graduate, while the failures are ruthlessly recycled. Only well-formed programs get to be tested in the rough and tumble Darwinism of the external world.

NOT RANDOM!

Note the generation of new subprograms and combinations of them is not a random event. See how difficult it is to shake oneself free of classical concepts. No, there will be relatively few combinations that have a Quantum Probability Form for them to fall into. To start, these QPF would have been provided by Nature just as molecular and atomic orbitals are provided by a beneficent Nature.

Later, there will be higher programs running that generate the QPF for the mixing and matching. This is exactly what our highly sophisticated immune system does in us. The immune system is capable of creating antibodies to millions of molecules only found in the chemists' test tubes. It is capable of generating antibodies to all but the simplest of the trillions of different molecules found in nature.

The immune system generates trillions of different such programs, each carried by a lymphocyte generated in the bone marrow.

Before it gets to run in the real world of the bloodstream looking for its complement to clutch and destroy, it first gets to run in a virtual environment created by a program running in the thymus. Only well-formed programmed lymphocytes are allowed back into the bloodstream as the mature T-cells of front-page fame. (The B-cells get tested in a different VR generated by an abdominal region called, for bird-related historical reasons, the bursa.) This is all well-documented and I am sure there was a great SciAm article about the it all recently.[1]

The virtual reality generated by the thymus is very simple. In essence, it just tests for one highly significant thing. The sequence goes like this

RUN a lymphocyte's linear program in the virtual reality
COMPLEMENT the QPF generated
WITH the QPF generated by all housekeeping programs also running
IF true
RUN destroy and recycle lymphocyte
ELSE
RUN activate and release on patrol.

Failures of this virtual testing is thought to result in the release of a lymphocyte that is programmed to attack a part of the body; arthritis is suspected to be such a malfunction. The lymphocyte is stimulated when it comes across cartilage, comple-

---

[1] True in any year, I am sure.

ments with it, and starts dividing, making lots of copies of the cartilage eating program. The resultant horde does its thing to the joints to great discomfort.

The Tetraplex

An even better example of testing in a virtual reality is the process of recombination between four strands of DNA in the mixing and matching and chromosomal rearrangement. This is closer in sophistication to the bacteria but, unfortunately, so little is known about speciation events that recombination is generally thought to be random even though there are well-known hot spots, as well as places strictly left alone.

We now propose that programs are run, and quantum probability forms get generated and tested in a virtual reality generated by the tetraplex stage.

While regular DNA involves just two strands of DNA, the tetraplex involves four strands. Surrounding these four intertwined condensed strands are a shroud of RNA and protein.

This internal testing of programs before they are released for external testing as new species explains why we just don't see lots of malformed individuals around. The external testing is the tip of the iceberg; most of the work has already been done internally.

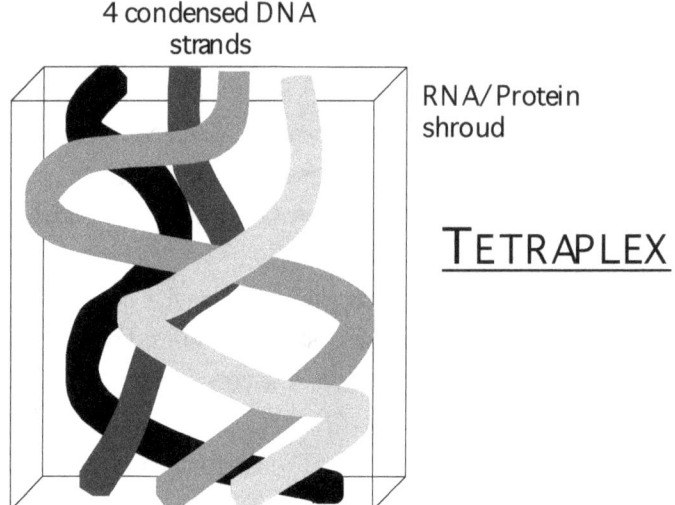

What are the implications for bacteria having developed a simple program that generates a virtual reality for program testing? Such a program would be just like Windows running on my Mac: while Windows thinks it is running on a real Intel chip, it's actually running in a virtual reality generated by a program, VirtualPC, running on Mac OS X.

We are talking about a simple and primitive VR generation, of course, for bacteria were perfected billions of years ago and have changed little since.

Such a virtual reality as a pre-testing environment could be expected to be useful in the current day. Do bacteria just blithely accept any old DNA that is passed to them by a kind stranger? I bet they don't. First, the programs carried by an incoming DNA gets run in the virtual reality to see how it fits in with the home team. If incompatible, the DNA is fragmented so that it's program is destroyed.

Bacterial viruses run programs that subvert this testing and so get to take over the bacteria. We can expect that the way a bacterium treats an incoming DNA from a pal, and the way a viral DNA bullies its way in, should delineate just what the virtual reality program looks like.

TOP DOWN, BOTTOM UP

Biology is currently using the bottom-up approach to understanding how life works. This is like describing TV as "light stimulates a sensor array to transmit a series of electrical impulses down wires to an antenna where they are imposed on a radio wave which is picked up and causes phosphors to sparkle in patterns on a screen."

This is all very true—it's even more complicated—but it gives no insight into the Super Bowl phenomenon, the "State of the Union" or even "Lucy."

Classical science, knowing nothing of probability forms, has to take the bottom-up approach. This is good. Complementing this, however, we need the top-down approach, deconstructing the programs that are running in living systems. I choose to call this approach dePrograming.

Multiplication

Then comes the external struggle for existence in the real world. How well does the program do when it gets a chance to actually run.

This is the Darwinian decimation that has been studied and documented so well by classical science. So, I will say no more about it.

Science and religion should take a time-out to really digest these new concepts of quantum probability before returning to the fray.

So we have now a simple picture that seems not unlikely: clay beds hoisting an early metabolism controlled indirectly by clay supersystems and their interactions.

All of this will depend on clay macromolecules being on the scene, of course. In the scenario outlined so far, we depend on there being a variety of clays around each with a specific set of capacities to provide the wavefunctions for carbohydrate manipulation.

So far we have imagined that each clay macromolecule assembled out of its monomers under the direct control of a internal system and the indirect control natural law setting the rules—in essence, the same way atoms and molecules originate out of their subsystems and then control them.

The problem here is that even if one particularly useful clay molecule emerges on the scene, just one system is not going to make much impact. And we might have to wait a long time for natural law to assemble that particular clay again as we can expect the internal systems of clay to be multitudinous and for many, many varieties to be equally possible.

TEMPLATE AND COMPLEMENT

The solution to this problem is remarkably simple. Understanding this takes no new concepts, thankfully. We have already seen how one system (such as clay) can provide the wavefunction for other systems (such as carbohydrate metabolism). All we have to postulate is that clay discovered what biologists might call the "alternation of generations."

We have already encountered the significance of patterns on the surfaces of macromolecules and how they interact. Clay has pattern-making capacity par excellence—they have plus and minus charge, H-bonding capacity, etc.

In exactly the same fashion that a clay surface can provide the wavefunctions for carbohydrate transformation, this pattern can provide the wavefunction for assembling a clay molecule with an exactly complementary pattern on its surface.

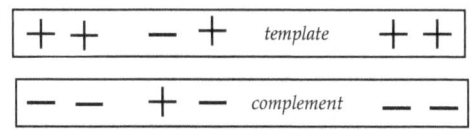

A biochemist would say that the first clay molecule acts as a template to produce its complement sequence.

Exponential growth

After the template and complement separate we have two possibilities:

the complement provides the wavefunction for a new template. We now have two template patterns on the scene—the template has multiplied.

the first template makes another complement—there are now two of them, complement multiplication.

The two of each can now repeat the cycle giving four of each, unlimited multiplication. And the power of exponential growth—which is what this doubling each cycle amounts to—is not to be underestimated. Any system which stumbles upon this simple method of multiplication via complements is clearly going to do well and become a major player. With just 300 such cycles of template-complement, for instance, a system could multiply to more that the number of particles in the known universe. Other factors of course—such as subsystem shortage—preclude such multiplication—but the possibilities are there.

CONSERVATIVE MULTIPLICATION

We can expect that, for any pattern-based system that has mastered the template-complement of multiplication, on Origin event is sufficient to make the system a major player. Once the very first of a qualitatively-different system emerges in an Origin event it can be rapidly multiplied by the template-complement process. The implication of this upward-compatibility process is that the template-complement process will be conserved up the hierarchy—its not going to change very much in history.

The providing of wavefunctions for running things exhibits enormous variation and "depth" of wavefunction hierarchies involved. The manipulative hierarchy involved in running reproduction—basically complex multiplication—remains remarkably unchanged all the way up. An example: The basic way we multiply human beings is exactly the same way single-cell plants and animals in pond water multiply: two haploid cells fuse, diploid cells multiply, diploid cell makes haploid cells, repeat each generation. The long-term world-line of the "human system" in history is just such a simple alternation of generations—the lineage diploid, haploid, diploid, haploid, etc. All the rest of the male-female dance is just a temporary housing that is built anew each generation.

As multiplication is so conservative we will only occasionally have to deal with any vertical movement up a hierarchy. The Origin of a new manipulative level of multiplication is, in almost all respects, the same as the Origins in the much more adventurous realm of manipulating stuff.

While I personally do not tend to the view that clay multiplication was that significant—I tend to think clay made buried lagoons of nucleotide-activated amino acids and nucleotides for the hell of it—this motif of multiplication via complementary patterns will appear over and over again.

Here we see a distinct quantitative difference between simple systems such a atoms which can only emerge by an origin process and pattern-based systems (such as clay perhaps) which can multiply via the complement process.

For atoms, the emergence of the first has little or no influence on the emergence of others. The rate at which a cooling plasma of electrons and protons forms hydrogen atoms is not influenced by the emergence of the first hydrogen—all of the atoms form under the direct control of internal systems provided by the indirect control of natural law.

For a system such as clay this is not true. While the emergence of the first is controlled by natural internal systems, the formation of more of the same can be controlled by the fixed internal systems, and the emergence of the first can have radical implications for the emergence of others.

Thus a clay molecule emerging originating in a rich supply of clay monomers could mop up the supply by multiplication—preventing any other clay molecule from originating. For the contingent stage of origins depends on there being subsystems around. Without them, the process stalls.

Fixation of a internal system through multiplication accomplishes two important things that are highly significant for the long-term survival of a system such as clay:

1. by fixing its internal system in its complement, the clay system frees itself from dependence on natural law to provide extra copies of itself Multiple copies do not need multiple origins. Just an Origin even suffices.

2. when multiplication replaces origins as the agent of change, the number of identical systems can increase exponentially. If one becomes two, then two become four, and four become eight, etc., then multiplication can rapidly dominate the scene. For instance, consider a clay molecule that can multiply once a day. While this might seem a slow rate of reproduc-

tion, this process could not go unchecked for even a year for, if each generation multiplied unchecked, the number of clay molecules by the end of the year would be more than the particles in the universe.

HORIZONTAL PROVIDING

We can now outline what wavefunctions are involved in multiplication. We will consider a pattern-based system able to manipulate its subsystems for long term stability and also to multiply itself—the providing of wavefunctions for the manipulation of subsystems and the providing of wavefunctions in the template-complement "alternation of generations."

The template and complement is the kind of change we have labeled horizontal, there is no movement in a hierarchy, the template and complement are on the same level. Multiplication is a horizontal, back and fore, mutual providing of wavefunctions.

In terms of somatic activity—basically body-building—most organisms have a "sense" strand template that is actively translated into protein and an "anti-sense" complement that is usually not translated. There are good mathematical reasons for this; and there are also many interesting exceptions.

Both template and complement are both capable of providing wavefunctions for other systems, they are, after all, basically the same—they are on the same horizontal level of sophistication of structure. But the wavefunctions they provide can be expected to be quite, quite different. For example, an all-plus clay molecule will have quite different catalytic activity to its all-minus complement pattern. The all-plus clay will excel at providing paths for minus-charged molecules—like amino acids—while the all-minus clay would excel at manipulating positive charged molecules like nucleotide bases. One way in which an all plus template could separate from its all minus complement after assembly is by allowing activated nucleotides and amino acids to assemble between the two strands, pushing them apart. As noted, separation is an essential, if sometimes neglected, aspect of multiplication. Perhaps clay supersystems fed by black smokers discovered that making activated amino acids and nucleotides was a Darwinian asset to being a long-term player.

Here the putative two strands of the clay are playing similar roles. The patterns that are being multiplied are basically the same as the wavefunctions being provided for structural advantage. This is a static, non-living type of situation we can suggest for something like clay. not the case in living systems which, as we will shortly see, does not involve static providing of wavefunctions; rather, it involves a flow of wavefunctions—one of the steps being a translation of a coded wavefunction into a provided wavefunction.

This step can be likened to what goes on at the lowest levels of the computer I am writing this thesis upon—currently a woefully out-of-date Mac. There the CPU is translating machine code—long strings of binary ones and zeros—into instructions which are executed. There is a limited repertoire of machine instructions—the instruction set for that chip—that just get run over and over again—millions of times a second. A computer program is a long, linear sequence of ones and zeros. Microsoft Word on this hard drive, for instance, is a string of them seven million from end to end. A machine code is equivalent to a triplet code in life. The sequence of bits is translated into a machine code which is executed. This is equivalent to the triplet code being translated into an amino acid which, as part of a protein, will be "executed" as it provides wavefunctions for molecular manipulations. While machine code is "strict"—each instruction is a specific sequence of bits—the triplet code is "degenerate" in that many triplet codes, up to six, can be translated as the same amino acid.

For all that, the translation process in both computers and life is very unforgiving. Even changing one item in the instruction set can have major impact of what comes out the other end—crashes and sickle cell disease are both unwelcome.

At the start of a computer program we might find a simple instruction that is translated into an instruction set such as "run the loading program at location 111111." This instruction is a string of binary digits, or bits, say: . The inverse of this is obtained by flipping all the ones into zeros and all the zeros into ones; another, quite different sequence of bits: . This is so different that one thing is certain: it is not an instruction, when translated, to "run the loading program at location 111111." It could be the instruction to "add zero: repeat."

So you would surely stump any programmer with the request for a word processor program whose inverse was a picture editing program. We are asking a programmer to be so clever that when we invert the seven million bits of Microsoft Word we get the seven million string that, when clicked upon, is Adobe Photoshop. Impossible! Perhaps something much, much simpler, is possible—but if so, I have never come across it in my reading.

This is why, in general, only one of the template-complement pair—the sense strand—does the work of coding for amino acids. The anti-sense strand provides the wavefunction for multiplication but usually has no other role to play.

An inert anti-sense is not always the case however; some viruses have taken compacting to such extremes that they have accomplished dual coding—both template and complement are translated into functional proteins. While they are only relatively simple proteins, this is extraordinarily difficult to imagine. But this is equivalent to what those minimalist viruses have done: the template encodes, say, a DNA ligase while the complement encodes, a protein coat.

Even our own species has exceptions to the sense-translated, anti-sense-not translated. "[The m-proteins were translated] genes that were embedded within an intron of the [NF- protein] gene. ... These 'genes within genes' carry their coding information on the DNA strand that is the anti-sense strand of the NF gene."[1] This is not quite as clever as the virus—the intron gets discarded before translation of the NF gene occurs—it is still quite unusual.

But, in general, only the sense strand is usually translated in living systems

Finally we come to the actual process by which the template-complements assemble upon, and then separate from, each other. The actual density of this in history, as always, will reflect probability density of collapsed wavefunctions. These are provided by natural law—one of our quantum operators—and any systems around in the environment that are capable of providing wavefunctions. While a clay molecule cannot influence the contribution of natural law, it can alter the systems in the environment.

To my mind, the ultimate in clay based metabolism might be the appearance of peptides that could manipulate fat metabolism. This would be difficult for clay itself, being so full of charge, as it cannot provide wavefunction for hydrophobic molecules.

With fats came the possibility of compartmentalization and a new type of combinatorial exploration. Certainty clay supersystems that could array themselves in lipids might have an advantage of stability to local water flow—a water-repellent raincoat perhaps. This could keep out the rough environment and allow delicate patterns time to assemble their complements. A somewhat porous coat, of course, as getting totally cut off is not a good idea.

While I doubt that such sophistication was attained by clay, the general pattern applies to all living systems. Just as there is a hierarchy of manipulation of stuff, there is a hierarchy of manipulation in multiplication. The manipulation hierarchy of multiplication is, however, much simpler; there are only a few levels. There is conservation and upwards compatibility. Human multiplication involves just three basic levels of manipulative ability

1. Manipulation and integration of DNA multiplication: Separation of DNA double helix. Assembly of complement on each strand to form double helix. Result; two helixes. Alternation of template complement. Only one strand is (usually) involved in providing wavefunctions for the structural side of things; the other strand is not transcribed. This is what bacteria basically do.

2. Manipulation and integration of cell multiplication. Extended chromosomes are condensed, a process that halts the provision of wavefunctions by the central genetic system. The cytoplasm is now running on auto-pilot and will do so until the chromosomes are unpacked and un-condensed at the end of the whole process. The wavefunctions provided by the re-awakened genetic system quickly bring things back under their beneficent control. The condensed chromosomes are paired up on the mitotic spindle and duplicated. There is now a foursome. Two of them are pulled one way by contraction of the mitotic spindle, the other two are pulled the other way. Each ends up with a chromosome pair, two daughter cells just

---

1 Christopher Wills, Exons, Introns and Talking Genes. Basic, NY (1991), p. 323

like the parent cell. Here we have an alternation between the paired and unpaired states. The condensed two-some and four-some states have the "complement" role here, they only have a brief time on the scene. Most of the history of chromosomes is in terms of the unpaired, opened-up state—the active, "template" role.

This is a sophisticated set-up unknown in bacteria and the difficulty in getting it up-and-running probably accounts for much of the billion years it took to get from bacteria to ameba. Once it was established it, just like the triplet code, took over the world: the mitotic spindles of animals, plants and fungi are essentially exactly the same.

In discussing the template-complement process we have neglected to take into account that, where wavefunctions are involved, we are always dealing with probabilities, probabilities that are rarely absolute certainties. In the process of multiplication via complements we have to take into account the probability of making mistakes in copying the pattern—alternating plus and minus forms—down the generations. While a template might arrange things so that the probability of its exact complement is very high, occasionally a low probability thing will happen—a sort of Murphy's Law of multiplication. This is an echo of the autonomy possessed by the subsystems. Remembering the inherent autonomy of systems to make choices, we should not be surprised if, occasionally, an unlikely pattern emerges that is not the perfect complement of the template.

Thus in our biogenetic clay Garden of Eden we can expect that the clay populations would be diverse in the extreme, but also related. As surface patterns can also be connect to the catalytic capacity of surfaces, we will see a corresponding variation in the metabolism thy indirectly control.

This can be compared with the champion multiplier, the DNA-based system that can make a complement with just one error in ten billion in general and much more in replicating crucial regions. Useful variation in higher organisms, useful in the sense of positive Darwinism, is rarely derived from mistakes in DNA. Such mistakes are rarely useful. Rather the variation so necessary to Darwinism is generated far up the hierarchy of genetic control in the process of recombination and chromosome manipulation.

A biologist would graph this variation as broad and wide for clay and narrow and sharp for a similar number of DNA generations of multiplication.

"The particular type of organization that exists in the dynamic interplay of the molecular parts of an organism, which I have called a morphogenetic or a developmental field [a internal system], is always engaged in making and remaking itself in life cycles and exploring its potential for generating new wholes."[1]

This is a sort of micro-origin, a small part of the system goes through an origin process while the rest is participating in a multiplication process. This is why the variation in a population of systems arising by mutation is called micro-evolution, a sort of hybrid multiplication-origin process.

Once a clay internal system has been fixed, natural law disappears from the picture—the displaced internal system of the catalyst indirectly controlling metabolism has been fixed. This is a primitive example of what we can consider a proto-genetic system involving both displacement and fixation—the emergence of a clay "gene."

As Dr. Cairns-Smith concluded: "Clearly there are further observational and experimental clarification to be made of the big question: Do mineral crystal genes exist? At this point I can only answer 'Quite possibly' and go on to the next question: Could mineral crystal genes evolve? The answer to this, it seems to me, is 'Yes, they could hardly help it.'"[2].

The choice of clay is not that significant. Other suggestions are proteinoid, iron sulfide, lipids, etc. Whatever it was, it must have had the basic attributes we find attractive in clay:

Natural abundance—the contingent factor in origin history

Catalytic activity capable of performing basic metabolism

---

1 Brian Goodwin, How The Leopard Changed Its Spots: The Evolution of Complexity Scribner's 1994, p. 176
2 A. G. Cairns-Smith, "The First Organisms," Scientific American 252, June 1985, p. 98

Plausible multiplication by the template-complement process.

We could extrapolate this perspective to the concept of clay supersystems controlling metabolic systems but it is difficult to see how clay supersystems could multiply by the template-complement process. Clay supersystems would be the original blob, just growing bigger and spreading their influence. Occasionally, some environmental upset might break a piece of which set up shop in some other location, but this is a pseudo-multiplication akin to crystals fragmenting—natural rules, not pattern rules are the ones in control.

Clearly multiplication does not increase the sophistication of the system—natural law is still just one step away—but it does enable the system to be a player, to be on the scene and explore the possibilities of positive Darwinism.

The ability of clay to multiply would have resulted in multiple copies of useful surfaces being on the scene—each with its entrained "useful" metabolism. Having multiple copies allows for rapid horizontal exploration of the possibilities of forming clay supersystems—the more systems doing the exploring the less time it takes for not-so-probable aggregations to check each other out.

A respectable mutation rate would provide for wide variation and for the clay systems to horizontally explore the possibilities of micro-origin.

Letting go

As always, the question of survival looms. There is what a biologist would call selection pressure, in an environment where many templates are competing for subsystems to make complements out of, any edge discovered by one will allow it to prosper. For positive Darwinism to occur we need a not-unlikely scenario for positive Darwinism to link keto-acid metabolic ability with the multiplication of clay. One plausible selection pressure on clay was what we might call the ability to let go.

Earlier on in the discussion we set the stage for our picture of multiplication by considering what happens when the complement forms on the template and then they separate. We neglected, at that point, to consider what would happen if the complement form but they did not separate. Clearly, not much. Separation is a key point in multiplication, for if the template and complement find each other's embrace so low-resistance that they never separate, the whole concept of multiplication stalls before it starts.

If clay complements tend to be sticky—the template and complement don't separate easily—this would be a bump-in-the-road for clay multiplication.

This is where keto-acids and simple sugars might have played a role. Much of clay pattern-bonding involves hydrogen bonds and, if sugars disrupt these bonds then they might facilitate separation of template and complement.

Such a multiplication-enhanced clay combo could multiply its components and their catalytic activity and monopolize the monomer resources provided by a beneficent natural metabolism. They would take over the clay bed eventually!

This is the most basic example of a living system imaginable: a clay supersystem with a metabolism supporting its multiplication—a example of a genetic-metabolic system that can multiply through time and space.

Another plausible suggestion is that the sugars helped the clay supersystems integrate, a glue to hold them together. And certainly polysaccharides are very sticky.

At this point we will just have to assume that there was a considerable advantage to keto-acid metabolism because, lacking a clay supersystem multiplication process, we are dependent on natural law to provide their internal systems, and this is not necessarily that likely. So we depend on the fact that once the capacity was established, it ensured the clay supersystems longevity for, on its continuance, depends the emergence of the amino acids and nucleotide bases. And we cannot realistically expect this clay process to be replaced until some sort of proteins and nucleic acids have appropriated the role of clay. The genetic takeover scenario championed by Cairns-Smith.

SEQUENTIAL METABOLISM

Once we can envision a not-unlikely scenario involving clay supersystems—and a similar will hold for any suggested starting surface—multiplying and manipulating simple carbohydrates—we can suggest a similar patter for the origin of clay supersystems with the capacity introduce the nitrogen atom and manipulate amino acids. We can expect that the pace of clay metabolism to be relatively slow and not that specific.

Here the horizontal exploration of the clay supersystems is being played out as a metabolic construct creating carbohydrates and amino acids and peptides. Peptides formed from amino acids have a wide range of properties and could reasonable be expected to have occasional helpful roles to play in clay life.

Organic molecules have many properties that would be useful to the survival and propagation of clay systems.[1] In his section "Organic chemistry without enzymes,"[2] Cairns provides a provocative view of how clay systems could develop the ability to manipulate organic. It is also established that the catalytic activity of metal ions can be made very specific in a structured environment.[3]

A similar thing can be envision for nucleotide base metabolism. The great benefit conferred by this ability could be expected in the manipulation the energy of phosphate bonds—the centrality of ATP today being the "fossil" remains—and polynucleotides to perhaps find a useful role as storage repositories of nucleotides.

Having monomers on the scene opens up the contingent possibility of stringing monomers into polymers. This is just another catalytic activity we could expect in clay—the ability, at some late date, to create relatively primitive examples of proteins and RNA. With primitive proteins and nucleic acids as products of proto-metabolism we have the contingent requirements for life, as we know it to emerge.

Perhaps the best argument against clay as a proto-metabolism is that clay is not involved at all and has left no fossil remnants except perhaps for the ubiquity of metal ions in protein and nucleic acid interactions. For the two founding macromolecules of current life are the proteins and the nucleic acids, not clays.

In the clay proto-metabolic-genetic system we allowed that clays have both a catalytic capacity and a template-complement capacity. Clay does both. In our kind of life, on the other hand, we see a division of labor:

a. Proteins excel in catalytic activity, in providing surfaces to setting the rules for molecular and macro-molecule transformations. The changes wrought by proteins are extraordinarily diverse and seemingly unlimited. Proteins, on the other hand, have zero capacity to multiply by the template-complement process.

b. Nucleic acids excel at the template-complement process. At its extreme, honed by selection pressures, its accuracy is such that errors are kept to below one-in-ten billion (admittedly with the help f many proteins). Nucleic acids, on the other hand make miserable catalysts (though they do have a small capacity to manipulate other nucleic acids).

Whatever it was that marked the transition from proto-metabolism to primitive life must have involved these two macromolecules discovering how to make up for the other's deficiencies; to be able to do together what clay can do alone, control metabolism and multiply.

The fact that RNA does have some catalytic function has prompted some to speculate that the pre-life was RNA without protein (let alone clay), and that all the catalytic manipulations were being performed by the RNA. "… it is possible under different reaction conditions to entice this [RNA] to act either as an RNA polymerase, endonuclease, ligase, kinase, acid phosphatase, or phosphotransferase. Thus many processes related to reproduction of the genetic information in a prebiotic RNA world could have been catalyzed by self-splicing [RNA]."[4]

---

1 Cairns-Smith, A. G., (1982). Genetic Takeover And The Mineral Origin Of Life Cambridge University Press, p. 308

2 Cairns-Smith, A. G., (1982). Genetic Takeover And The Mineral Origin Of Life Cambridge University Press. p. 310

3 T. J. McMurry, K. N. Raymond, P. Smith, "Molecular Recognition and Metal Ion Template Synthesis," Science, 244, 1989, p. 943.

4 P. A. Sharp & D. Eisenberg, "The Evolution of Catalytic Function," Science 238, 1987 p. 729-730, 807

Operating System of Life

We have now dealt, in broad terms, with the evolution of the 10,000 or so programs in a typical bacterium.

All of this, however, depends on the one basic operating system, the ribosome, to actually run the programs and allow them to generate quantum probability forms to contribute to the composite.

The question now becomes, How did the operating systems originate? How did they evolve?

There are remarkably few operating systems for us to consider—for the triplet code, RNA system is just the first of the few we need to look at.

The evolution of operating systems is Macroevolution. For when a new operating system emerges on the scene, a whole new realm of possibilities opens up and exploration is exponential. This period of top-level experimentation ends, however, when they become subprograms and get called a lot by other programs. Change quickly becomes impossible, and that stage of evolutionary exploration is ended.

There are just four macro-evolutionary events. The chart also gives examples of when this is the 'top level' of programming sophistication. Otherwise, the level becomes a relatively invariable subprogram for a higher level.

Operating system	Linear program	Virtual reality
Basic OS	triplet code RNA	bacteria
Cell OS	spindle RNA	yeast
Organ OS	virish RNA	plant
Nervous OS	glial RNA	Fish mind
Emotional OS	basal RNA	reptile mind
Symbolic OS	bellum RNA	ape mind
Spirit OS	dual RNA	human mind

Virtual Reality

Each of these OS at the peak of innovation and exploration can be expected to develop programs that generate a virtual reality. A VR in which other programs can be run virtually; they can be tested internally before they are let loose in the external world. Much of the following is pure speculation, but I hope you enjoy it.

Again, this is like my Mac. The OS X runs many programs at the same time in threads that get a certain amount of CPU time. In the following screen grab, there is a program called "null" taking up a lot of time. This is the Mac OS 9.3, a virtual environment in which I can run my old programs. To OS 9, the virtual environment is a virtual CPU that it is running. If OS 9 gets crashed, OS X will burp, inform me, and blithely continue running the real CPU. On a real machine, such an OS 9 crash would necessitate a total restart.

Process ID	Process Name	% CPU	# Threads	Real Memory	Virtual Memory
311	Safari	0.40	6	54.96 MB	244.98 MB
175	WindowServer	3.40	2	53.79 MB	217.07 MB
585	(null)	91.10	14	39.79 MB	1.13 GB
587	Backup	0.00	4	20.24 MB	148.99 MB
281	MenuStrip	0.40	5	15.15 MB	158.53 MB
578	Activity Monitor	4.40	2	13.87 MB	165.23 MB
177	ATSServer	0.00	2	12.36 MB	87.39 MB
471	Stickies	0.00	1	12.36 MB	153.82 MB
588	BackupHelper	0.00	1	10.50 MB	36.73 MB
205	Finder	0.00	1	10.42 MB	157.11 MB
582	Grab	0.40	3	8.71 MB	161.20 MB
581	Windows Media Player	0.90	5	6.47 MB	152.18 MB
204	SystemUIServer	0.00	1	6.42 MB	148.99 MB
184	loginwindow	0.00	4	4.73 MB	123.52 MB
274	Mirror Agent	0.00	9	3.70 MB	145.43 MB
203	Dock	0.00	2	3.49 MB	138.98 MB
279	iCalAlarmScheduler	0.00	1	2.93 MB	133.51 MB
199	pbs	0.00	2	1.61 MB	44.41 MB
280	iTunes Helper	0.00	1	1.30 MB	124.53 MB

Basic OS virtual reality—Virtual ribosomes on which programs can be run virtually. Foreign DNA gets tested first.

Cell OS virtual reality—Virtual DNA in the nucleus mixed and matched. Recombination patterns tested first here, then expressed on real chromosomes.

Organ OS virtual reality—Virtual chromosomes in the nucleus mixed and matched. Recombination patterns tested first here, then expressed on real chromosomes.

Nervous OS virtual reality—Virtual neuronal patterns tested first in the virtual reality generated by the glial cells. Well-formed programs are passed to real neurons.

Emotional OS virtual reality—Virtual emotional patterns tested first in the virtual reality generated by the basal ganglia. Well-formed programs are passed to the upper brain for real action.

Symbolic OS virtual reality—Virtual symbolic programs tested first in the virtual reality generated by the cerebellum before being passed to the front brain for real action.

Human mind OS virtual reality—Virtual symbolic programs run in a virtual reality. I can only suppose that this virtual reality is the one that I inhabit inside my head. Similar to the VR where, in your mind, dear reader, you are tossing around these words and ideas to see if they make any sense.

Basic OS 1.0

Now we will look at the various operating systems on which the RNA programs are running, starting with the simplest, and best characterized.

The most basic is the bacterial-organelle operating system. This is the basic triplet code method of protein synthesis that is being thoroughly explored. We will refer to it as the basic operating system or BOS.

Ever since BOS 1.0 was released, it is has remained virtually constant for billions of years (unlike the Mac operating systems I have followed from 1.0 to OS X.) The operating system in the microbe that turns milk into yogurt is exactly the same as the one running inside my mitochondria.

Next, a program and an operating system need a processor to run on. My PowerBook has one; my office G4 has two. A processor can only run one program at a time—my Mac OSX evades this bottleneck by running dozens of 'threads' but it can only deal with them one at a time so it spends milliseconds running each one in turn.

The bacterial equivalent of the processor is the ribosome; this is the BOS for the bacteria/organelle programs to run on.

BASIC OS
RNA linear program runs via protein
Generates Quantum Probability Form
Stuff falls into collapsed probability form

The number of processors in a single bacterium, however, runs to the hundreds of thousands.

As anyone who works with computers will attest, changing operating systems is a major hassle. Nothing old runs anymore; new versions have to be purchased. Living systems are fortunate not to be cursed with this "new, improved" burden. At last, we have the first stage of quantum life science, the Basic OS.

Cell OS 1.0

The operating system that runs cells—plants and animals—has elements similar to the bacterial level—there is a more sophisticated version of the ribosome, for instance.

Many programs, however, seem to be also running on the cytoskeleton network which is capable of remarkable expressions when properly programmed. The most dramatic example of this is the spindle of cell division that generates probability forms for the chromosomes to fall into. The units seem to be a dozen-or-so proteins, such as the actins, that pop together like legos.

Yet, another is the spliceosome system that snips all the non-triplet code out of DNA-to-RNA transcripts before being sent out of the nucleus. They are complexes of RNA and proteins, and an area of the nucleus, the nucleolus, seems to an element of the operating system here.

In the 1950s, the greatest advance in understanding the mechanisms of life was the elucidation of the triplet code in the DNA. This mapped the sequence of bases on the DNA to the aminoacid sequence in proteins. It only emerged much later that only a fraction of the information encoded in the DNA of complex organisms ever makes it out of the nucleus and gets translated by the ribosome into protein aminoacid sequences.

Molecular geneticists have found that, although the DNA is transcribed into mRNA, long lengths of the information are neatly and precisely excised from the mRNA before it is transported out of the nucleus. The DNA sequences that are excised are the "introns"—and they can be hundreds, even thousands, of bases long—while the remaining sequences are the "exons." It is only the exons that are spliced together by spliceosome complexes and transported out of the nucleus to direct the assembly of protein in the cytoplasm ribosomes. Having exons—roughly equivalent to protein domains—separated by introns has the advantage of being able to shuffle bits of proteins around and make new multi-functional proteins.

The mechanism separating the intron and exon material involves small complexes of RNA and a variety of proteins. They are called small nuclear ribonucleoproteins (snRNP).

"There are many different kinds of snRNP's, and functions have been assigned to only a few. ... They are the critical components of a sophisticated molecular assemblage called a spliceosome. As such, they take part in the splicing of mRNA ... a delicate operation that must be carried out with the utmost delicacy and precision.. Perhaps it is not surprising, then, that the snRNP's in spliceosomes specialize: each performs a different task during the splicing procedure. The picture of snRNP's working in concert in the spliceosome suggests nothing if not a well-oiled machine. ... One of the most intriguing aspects of spliceosome function is that the entire assemblage, rather than any individual component, seems to be responsible for the catalysis of the splicing reaction."[1]

"Sequence families similar to **Alu** are characteristic of mammals. They are not known to contribute to the survival of the organism. ... However, their presence does have important effects on mammalian evolution because interactions between **Alu** sequences at different sites may cause structural rearrangements of the chromosomes. The rapid evolution of chromosome structure in mammals may therefore be caused by the presence of **Alu**-like dispersed sequences. ... In the grasses, much of the DNA consists of short sequences, with copy numbers that may be greater than a million, arranged in

---

[1] Joan Argetsinger Steitz, "Snurps," Scientific American, June 88, pp. 56-63.

tandem blocks, distributed over the chromosomes, but concentrated in certain regions. ... Highly repetitive DNA of this kind occurs throughout the animal and plant kingdom, but in varying amounts."[1]

While current science has decoded the triplet code—how the monotonous combinations of just four "letters" A, T, C, G codes for protein—it has yet to decode the patterns of this seemingly useless DNA.

From the perspective we have developed, we can expect that this DNA contains all sorts of "meaningful" codes

Genetic Programs

Life, it would seem, excels at the manipulation of quantum probabilities. What about a cell? The genetic system is running many, many programs that are generating a plethora of quantum probability forms that get filled in by all the stuff. A healthy cell is one where all the constituents are in a high-probability state. Disruption and disease moves things to an improbable state; healing occurs as the stuff falls back into the highly probable state.

One nice thing about this perspective is that it clears up a question I came upon in high school. The atoms in my body are all replaced in about a week or some shockingly-short period. What remains constant then? I wondered. Now, at least, I have a hint of answer—a quantum probability form that stays relatively constant.

Could development be a filling in of quantum probability forms as they are sequentially generated by an RNA-borne linear program? Interesting support for this is the quantum prediction of "fitting into a form" in two ways. While the great majority of people have their heart and lungs fit into the chest with the heart on the right, a few have everything reversed. But it all fits perfectly and such people have no problem. Very occasionally, a person has all the internal organs in the flipped position. Again, the fit is perfect. While such flexibility is built into the quantum view, it is alien to the classical perspective.

The closest classical analogy to a cell would be a computer running a factory producing computers programmed to run the computer factory. As this ascribes a Godly position to Bill Gates and Steve Jobs, and they don't need more elevation, we will pursue the analogy no further.

SELFISH DNA

What is all the intron DNA used for if its information never gets to make protein? Most scientists have little to say about this strange surplus of DNA, although Richard Dawkins has been a little more inventive. He theorized that science has it all back to front: A body is actually only DNA's way of making more DNA and that much of the DNA had no function and in the triplet code was gibberish. This 'hanger-on' DNA Dawkins called 'selfish DNA.' Once selfish DNA had established itself, it just replicated itself down the generations along with the DNA doing the useful work[2].

In the genetic model used in the modern evolutionary theory, there is only one type of information in the DNA, and that is stored in exon DNA: aminoacid sequences written in the triplet code. We are proposing, however, that there is a great deal of information in DNA that has nothing to do with aminoacid sequence directly. The intron DNA provides a possible resting place for such information that is not translated by the triplet code, ribosomal mechanism.

"Because a direct function for this DNA is not readily apparent, it is often disregarded. However, a substantial portion of this excess DNA may specify genetic and structural partitions and may also provide essential recognition features that are important for orderly gene function. ...Indeed, excess DNA may be essential for the efficient 'compartmentalization' of genes at several hierarchical levels of organization."[3]

This concept that information is stored in the DNA in ways other than the Triplet Code is well supported by recent work on the effects that nucleic acids can have directly on other nucleic acid and their function.

---

1 Maynard Smith, John, Evolutionary Genetics, Oxford University Press, 1989, pp. 217-221.

2 Dawkins, R., (1976) The Selfish Gene, Oxford University Press.

3 L. Manuelidis, 1990, "A View of Interphase Chromosomes" Science 250 pp. 1533-1540.

One review of these developments stated that "Among the best studied ... are the self splicing introns ... [This self splicing intron] can act as either as a RNA polymerase, endonuclease, ligase, kinase, acid phosphatase, or phosphotransferase."[4]

A DNA sequence, it seems, has ways of controlling other DNA that does not involve it being transcribed into an aminoacid sequence. The authors of the same review think that this is good evidence for active pre-biotic RNA, a possibility discussed in Dr. Cairns' argument for the low-tech role of clay—see the Appendix.

It has already been shown that intron DNA has a specific (non-triplet code) vocabulary, "The presence of idiosyncratic words implies that the primary structure of introns is far from being random. We conclude that introns do carry some messages and, hence, should not be regarded as 'nonsense' DNA."[2]

Another possible mechanism has been raised by the recent work being done on the methylation of DNA. "Methylation of DNA is a ubiquitous phenomenon ... In eukaryotes, there are no established functions for DNA methylation, though recent evidence suggests that it may regulate gene expression.[3]

Another possibility is the actual structure of the nucleus itself and its control of the genetic information. For example, the nuclear matrix (the insoluble structural framework) has been shown to be involved in the splicing of introns and exons.[4]

Regulation mechanisms can be expected to be heavily involved in the process of evolution of the higher organisms. As one worker put it, "The most prominent evolutionary mechanisms in prokaryotes involve mutation and other genetic operations involving the sequence variability of DNA. The differences within wide taxonomic categories of metazoans are often regulatory ... The major adaptive radiations among these forms are likely to have been mediated by regulatory evolution."[5]

If it is correct to say that intron DNA has a host of regulatory functions, we can also conclude that the evolutionary development of the more sophisticated levels of regulation will involve intron DNA (regulation) rather than exon DNA (metabolism).

Whatever the details, however, the point has been made: There is plenty of room in the DNA for storing codes other than the storage of sequence information in the triplet code.

"It has been shown that the SMN protein is involved in spliceosome biogenesis and pre-mRNA splicing, there is increasing evidence indicating that SMN may also perform important functions in the nucleolus... These studies raise the possibility that SMN may serve a function in rRNA maturation/ribosome synthesis similar to its role in spliceosome biogenesis."[6]

Both operating systems and their codes are not well characterized and are currently under intense investigation. But it would clearly behoove geneticists to learn all about massively-parallel computer programming and do top-down studies to complement the well-established bottom-up approach currently doing so well.

The eukaryote ribosome and the protein-legos then are elements of the cell operating system. An active cell can have millions of eukaryote ribosomes in the cytoplasm (along with the myriads of prokaryote ribosomes in the mitochondria) and millions of the lego proteins. Massive processing running relatively few programs.

---

1 Sharp, P. A. and Eisenberg, D., "The Evolution of Catalytic Function" Science238, pp. 729-730, 807.

2 Beckmann, J. S., Brendel, V., and Trifonov, E. T. (1986) "Intervening Sequences Exhibit Distinct Vocabulary" J. Biomolecular Structure & Dynamics 4, pp. 391- 400.

3 Essani, K., Goorha, R. and Granoff, A. (1987) Virology 161, p. 211 - 217

4 Zeitlin, S., Parent, A., Silverstein, S. and Efstratiadis, A. (1987) "Pre-mRNA Splicing and the Nuclear Matrix" Molecular and Cellular Biology 7, p. 111-120

5 Conrad, M., Brahmachari, S. K., and Sasisekharan, V., (1986) "DNA Structural Variability as A Factor in Gene Expression And Evolution" Biosystems 19 pp. 123-126.

6 Wehner KA, et al, Brain Res. 2002 Aug 2; 945(2) pp. 160-73.

The ribosomes in our type of cell are similar to those in a bacteria, just a step up in size and sophistication. Much is still mysterious about how the cytoskeleton is organized. An organelle called the centriole seems to play a central role—it certainly does in cell division where its two aster-poles separate into the spindle which generates a quantum probability form that separates the chromosomes.

What could be programming the centriole and cytoskeleton? We need something that can carry a linear program to the centriole. It is messenger RNA (mRNA) that carries a program to the ribosomes.

Let's look at the usual suspects. Could it be RNA that is programming the centriole? I googled "centrioles and RNA" and right there at the top I found this:

"Evidence for a functional role of RNA in centrioles… We conclude first, that centrioles contain RNA which is required for initiation of aster formation, and second, that the centriole activity or ability to assemble a mitotic aster is separable from the basal body activity, or ability to serve directly as a template for microtubule growth."[1]

Sounds like a 'yes' to me. Spindle RNA (spinRNA) will not be carrying a program in the triplet code—the ASCII of the living world—but written in another code. This 'spinlet code' will be linear and, just like the triplet code and the ASCII, it will involve patterns.

Now it just so happens that our chromosomes—animals and plants alike—contain vast stretches of DNA patterns that are total nonsense in the triplet code. Like thousands of repeats of sequences of nucleotides such as **AAGGCCT** over and over again. This 95% or so of the genome has been called a variety of names, none polite, such as selfish and junk.

One reason for this disdain is that they apparently are not transcribed in RNA and shuttled out of the nucleus into the cytoplasm.

I say apparently: There are hundreds of thousands of ribosomes and hundreds of thousands of tripRNAs pouring out of the nucleus to program them. But there is only one centriole. So, we might expect the proportion of spinRNA to mRNA to be about one in a million; probably below the resolution of current techniques.

Where are these few, non-triplet code patterns on the spinRNA coming from, I wonder. Perhaps sections of the junk DNA are transcribed into small amounts of spinRNA that conveys a program to run on the centriole.

As the spindle forms and separates in its majestic and stately fashion, we can expect that a maximum number of programs are being sent sequentially to the centriole/asters/spindle. So, this is where the maximum of junk transcription will occur if the suggestion is correct.

There is just one cell operation system in release—Cell OS 1.0—the ribosomes and mitotic spindles in my skin cells are identical to those in an oak's bark.

Just as a bacterium can jump from the active to the spore state, so do the much larger cells of our kind.

Unlike bacteria, cells have triplet code genes fragmented into exons, spaced apart by introns that are non-triplet code and are not translated into protein.

While all cells have such fragmentation, it appears to increase with overall sophistication. This also allows for a great deal of mixing and matching of the exon modules so that quite different proteins get made which generate quite different QPF.

A recent overview states it as:

"In the {simplest cells} the splicing machinery can recognize only intronic sequences of fewer than 500 nucleotides, which works fine for yeast because it has very few introns, averaging just 270 nucleotides long. Bust as genomes expanded during evolution, their intronic stretches multiplied and grew, a cellular splicing machinery… [was] forced to switch… to a

---

1 Heidemann SR, Sander G, Kirschner MW. Cell. 1977 Mar; 10(3): pp. 337-50.

system that recognized short exons amid a sea of introns. The average human protein gene is 28,000 nucleotides long, with 8.8 exons separated by 7.8 introns. The exons are relatively short, usually about 120 nucleotides sequences [40 aminoacids when translated], whereas the introns can range from 100 to 100,000 nucleotides long.

"The size and quantity of human introns—we have the highest number of introns per gene of any organism—raises an interesting issue. Introns are very expensive habits for us to maintain. A large fraction of the energy we consume every day is devoted to the maintenance and repair of introns in their DNA form, transcribing the pre-mRNA and removing the introns, and even to the breakdown down of the introns at the end of splicing reaction... By generating more than one type of mRNA and, therefore, more than one protein per gene, alternative splicing certainly allows humans to manufacture more than 90,000 proteins without having to maintain 90,000 genes... Our genome already contains some 1.4 million ALU copies, and many of these ALU elements are continuing to multiply and insert themselves in new locations in the genome at a rate of about one new insertion per every 100 to 200 births. The ALUs were long considered nothing more than genomic garbage, but they began to get a little respect... Thus, ALU sequences have the potential to continue to greatly enrich the stock of meaningful genetic information available..."[1]

Sounds like encoded information is being passed down the generations.

Naturally, the classically-trained geneticists who wrote this consider the sprinkling of ALUs as random. We can expect, however, that the ALUs end up where they are because of a program-quantum probability form. Their positioning is specific and obeys a programming syntax.

This programming is clearly very important as, "Almost half the human genome is made up of transposable elements, ALUs being the most abundant."[2]

As we shall see, there is plenty of use that all this RNA can be involved in as layer upon layer of code is laid down.

Cell OS
RNA linear program runs
Generates Quantum Probability Form
Stuff falls into collapsed probability form **BASIC OS**

Organ OS 1.0

Next up the hierarchy of sophistication is the operation system of organs. How does our liver organize itself, how does a corn plant?

Let us assume that what we have seen so far still holds true on this level (nature is very conservative at the basic levels):

The organ is a consequence of a quantum probability form generated by programs running on a conserved operating system. The programs are linear and few; the processors multitudinous.

The obvious processors are the cells. Each of my organs has trillions of them in about 200 or so varieties.

What could be carrying linear programs to the cell processors?

A clue is perhaps to be found in the AIDS epidemic. HIV is a strand of RNA wrapped in a protein coat. The RNA is not very long as such things go.

When this HIV RNA enters a T-cell (the T stands for thymus-trained, a B-cell is trained in the 'bursa' region) the HIV program does a remarkable thing: it flips the considerably more massive cell from its healthy state to that of being an HIV factory. One single strand of RNA can suborn a whole cell. All viruses, both the DNA and RNA variety, behave in this way. The calcium ion and the prions have a similar effect in their own provinces of action.

---

1 Gil Ast, The Alternative Genome, Scientific American, April 2005, p.63.

2 Gil Ast, The Alternative Genome, Scientific American, April 2005, p. 64.

When HIV-RNA enters a T-cell, the first thing it does is get transcribed by a ribosome into an aminoacid chain, which folds, the assembly code runs, and generates a QPF with one special ability. It copies HIV-RNA into the human DNA that prup with a ribosome and get transcribed into

I have always thought it stupid of human cells to be so vulnerable—just one rogue strand of RNA, added to the millions already in there, causes a massive jump in state to occur. Why so defenseless?

Jumping to the conclusion: perhaps cells pass RNA to each other, this orgRNA carrying a linear program that runs in the cell. (I discount DNA, as all signs are that it is religiously segregated from the cytoplasm except during division.)

This is what the rogue HIV does, after all. Rather suggestively, our chromosomes are riddled with tens of thousands of genes for reverse transcriptase in various states of disrepair. And this is the very enzyme that allows the HIV to suborn a cell. All very indicative, you must admit.

This was second on the list when I googled "transfer of RNA between cells:" Evidence for transfer of macromolecular RNA between mammalian cells in culture.[1]

Unfortunately, that's all I could pull up, but it suggests that RNA might just be involved, yet again. (How many RNA codes are there? Clearly, we are still living in an RNA world!)

A cell in a developing organism receives a program on an orgRNA, obediently runs the received program, and generates a new QPF. The filling in of this new QPF is called differentiation. The sender of the RNA being in the position of User to the cell program. Just like the roles of bacterium and environment in sporulation as earlier discussed.

Just how does HIV suborn a cell? Clearly more than a simple protein or revamping of the cytoskeleton is involved. A new Master Program is being run on the cell OS and the infected cell turns into a virus factory.

We can expect that the code is not the triplet code, and is not the spinlet code: it can be called the organ-let code. One of the instructions will be to pass on some orgRNA to the cells around it. This can be modified to carry a result along with it. For in programming, counters are very important. Regular programming abounds in statements in the linear progression such as:

>   FOR N = 1 TO 10
>   RUN program
>   NEXT N

A changing count could easily be kept on RNA as it passes from cell to cell by something as simple as the number of a repeating sequence. The telomeres seem to keep a similar count of cell division.

This following might represent a simple counter:

>   FOR T = 5 TO 1 RUN NEXT T

Turning Worms

A similar thing apparently happened during evolution and we are living with the consequences. For all animals, the first stage of development is the formation of a hollow ball of cells with a hole in it. This first hole is the mouth; a second hole then forms as the anus.

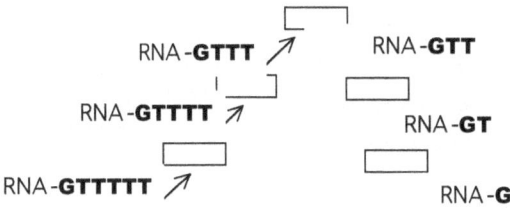

This pattern holds for a wide variety of "primitive" animals, including the hugely successful simple worms and complex insects.

In the lineage that leads to fish, frogs, dinosaurs, elephants and us, however, there is a sudden flip in the developmental process: the first-hole mouth-to-be flips to being the anus; the second-hole anus flips to become the mouth end.

---

[1] Kolodny GM., Exp Cell Res. (1971, Apr) 65(2) pp. 313-24.

While this is hard to reconcile with classical concepts, it is perhaps an example of there-are-always-two-ways-to-fit into a probability form. Something about this flip of the mouth to the second, and perhaps, more sophisticated hole opened up a world of possibilities including our big brains.

DEVELOPMENT

In fact, it is clear that RNA mediated, HIV-like reverse-transcriptase transformation of the DNA has been important in the history of our evolution:

"The commonest of all the [retrotranspons, long considered a genetic parasite] is a sequence of 'letters' known as LINE-1. This is a 'paragraph' of DNA between a thousand and six thousand 'letters' long, that includes a complete recipe for reverse transcriptase near the middle. LINE-1s are not only very common—there may be 100,000 copies of them in each copy of your genome—but they are also gregarious, so that the paragraph may n=be repeated several times in succession on the chromosome [a la Hox]. They account for a staggering 14.6% of the entire genome, that is, they are nearly five times as common as 'proper' [triplet code] genes The implications of this are terrifying. LINE-1s have their own return ticket. A single LINE-1 can get itself transcribed, make its own reverse transcriptase, use that reverse transcriptase to make a DNA copy of itself and insert that copy anywhere among the genes...

"If LINE-1s are about, they too can be parasitized [it is supposed] by sequences that drop the reverse transcriptase gene and use the one in the LINE-1s. Even commoner are shorter 'paragraphs' called ALU. Each **ALU** contains between 180 and 280 'letters', and seems to be especially good at using other people's reverse transcriptase to get itself duplicated. The **ALU** text may be repeated a million times in the human genome—amounting to perhaps 10% of the 'book.'

"For reasons that are not entirely clear, the typical ALU sequence bears a close resemblance to a real gene, the gene for a part of the... ribosome. This gene, usually, has an internal promoter, meaning that the message 'READ ME' is written in a sequence in the middle of the gene...

"The genome is littered, one might almost say clogged, with the equivalent of computer viruses... Approximately thirty-five percent of the human DNA consists of various forms of [viral-like] DNA, which means that replicating our genes takes thirty-five percent more energy than it need. Our genome [and this we have to disagree with] badly needs worming.

"There are sequences even shorter than **ALU** that also accumulate in vast, repetitive stutters... The 'word' can vary with the location [on the chromosome] and the individual, but it usually contains sentences of the same central 'letters': **GGGCAGGAXG**... The significance of this sequence is that it is very similar to one that is used by bacteria to initiate the swapping of genes with other bacteria of the same species, and it seems to be involved in the encouragement of gene swapping between chromosomes in us as well...

"It turns out that the repeat number is so variable that everyone has a unique genetic fingerprint; a string of black marks looking just like a bar code."[1]

Sounds like info is being passed down a lineage to me.

Processors are the cells. Each is running a program depending on the RNA received from the neighbors. Running this program generates a QPF. Each cell in the organ contributes this to the composite whole, cell-level QPF.

Just like Mandelbrot, just like bacteria, this internal composite probability amplitude, when collapsed, has a quantum probability form to it that gets filled in by stuff. When filled in, this composite QPF is what we call a healthy cell. While the OS remains constant, the program running can change as easily as a ribosome can start translating another mRNA when finished with the first.

In a healthy organ, while each cell constantly receives a program to run, it is the same program that arrives. It keeps doing what it is doing.

---

[1] Ridley, M, Genome, Harper Collins, 1999, pp. 125-132.

When something like healing is called for, however, a new message arrives and the cell switches to an aggressive division mode appropriate for healing.

The arrival of a different RNA linear program can change many things at the analog end results: cell adhesion, cell division, cell differentiation, cell death, just a few dozen at most.

As the stuff fills in the quantum probability gradients being generated by running RNA programs, we observe the healthy functioning of a mature liver, or the rapid healing of a damaged lobe as the QPF change.

Cells falling into probability forms generated by programs running in the cell processor. A liver has a dozen or so types of cells—different processors—and trillions of copies of each. Truly massive parallelism.

There is also patterning on the DNA passed down a lineage. An example is methylation which adds lots of oily spots all over the DNA. This is known, for instance, to signal if it was Mom or Dad who contributed that chromosome. Most methylation lies within transposons such as ALU and LINE-1.[1]

"…the first [Hox] genes defined the head end of the fly and the last [Hox] genes made the rear end of the fly They were all laid out on order along the chromosome—without exception.

In mice, there it was again: almost the same 180-letter string—the homeobox. Not only that, the mouse turned out to have clusters of Hox genes (four of them, rather than one [in the fly]) and, in the same way as the fruit fly, the genes in the clusters were laid out end-to-end with the head genes first and the tail genes last… What was doubly strange was that the mouse genes were recognizably the same genes as the fruitfly genes… By having four [Hox cluster}, we and the mice have rather more subtle control over the development of our bodies than flies do with just one Hox cluster."[2]

Organ OS
RNA linear program runs
Generates Quantum Probability Form
Stuff falls into collapsed probability form **Cell OS**
BASIC OS

---

1 Ridley, M, Genome, Harper Collins, 1999, p. 182.

2 Ridley, M, Genome, Harper Collins, 1999, p. 177.

Eden, Wombs and Beds

We have explored the internal aspect of quantum probability so far. But, as mentioned earlier, every quantum calculation ends with the collapse of the wavefunction and the filling in of probability forms.

This is the contingent nature of history so well-explicated by the late Steven J. Gould. In brief, before stuff can fall into a quantum probability form, the stuff has to be already on the scene. In order for a quantum probability form to be filled, all the ingredients have to be present.

Our earlier example was of a protein embracing a calcium and jumping to fill a quite different QPF. In the absence of calcium, that QPF is empty, it is not filled in, and can play no role in the external world of interaction and the sharing-exchange of bits of self, the stuff filling in one's QPF. (Apparently, good car drivers extend their sense of self to embrace their car, sometimes even emotionally.

The great leaps in evolution each involve the emergence of a new, more sophisticated operating system. These leaps are few in number and took the longest times.

We can expect that the twin pinnacle of programming possibilities at each level of language involve:

First, the emergence of a VR in which programs can be virtually tested. This is the explosion of innovation each level went through as witnessed in the historical record.

Second, the elaboration of all the ingredients for the next level of programming language and a more sophisticated operating system for it to run on. When all the ingredients are together, they can jump to the new configuration of the QPF provided by Nature. We can call the last ingredient to appear on the scene, thus setting the scene for the quantum jump, the calcium factor. The jump is the same, just on a different scale.

We can take it that a small bacteria is about the maximum size over which natural quantum jumping occurs for it is rare for active components in living systems to be larger than this size.

Womb Eden 1

It took the bacteria and their colonies about a billion years to generate all the bits needed to fill in the QPF of the prokaryote type of cells.

When all the ingredients were in the right place, they all quickly popped into the now-highly probable configuration of the cell operating system.

The place was probably in a stromatolite and the eukaryote cell was rapidly perfected from this earlier prototype.

Stromatolites

The fossil evidence for the establishment of sophisticated bacterial colonies is striking in the billion-year old stromatolites. Similar to ones extant in a few exotic locations today, these are stratified layers of many different single-cell organisms. The strata reflect the history of their sequential evolution. These stromatolites are, "A dome-shaped structure about a foot high ... made of hundreds of wafer-thin layers of rock [whose] counterparts exist today ... in shallow water in restricted locations, such as the coast of Australia."[1] These are formed today by a primitive type of algae and the oldest of these structures found so far seem to be about 3,500 million years old.

The topmost level is the photosynthesizer. The ability to use light as an energy source—forever liberating life from proto-metabolism—seems to have been discovered early on. It is apparently the result of a duplication of capacities.

First to be established was the ability of chlorophyll molecules—complex but not too difficult to make—to absorb light energy and raise hydrogen to an energy level where they can be used to create the two basic coenzymes needed for biosynthesis: reducing power in the form of NADPH and energy in the form of ATP.

While the formation of ATP by light energy is cyclical—the H-ion is returned back to the start—the formation of NADPH is not since the H is lost from the system and has to be replaced from somewhere. Unfortunately, there are few ready sources of H at the energy level required for this process to work. As shall see, this problem was quickly solved.

Chlorophyll works because the wavefunction inhabited by an unexcited electron is spatially very different from that inhabited by the excited electron. The excited electron is snatched up by a bucket brigade of small molecules and its energy can be used to drive NADPH and ATP synthesis.

With these two, the fixation of carbon dioxide becomes possible, an endless supply of carbon opened up.

The basic reaction of this first step in photosynthesis is to create glucose out of carbon dioxide by reducing it with the NADPH driven by the energy released from ATP.

The problem of finding a source of the H atoms was solved in a simple manner by duplicating the photosynthetic apparatus. This second program-in-action specialized in taking the H atoms from water—at a low energy—and raising them up to the level required by the original system.

Only the hydrogen depleting synthesis of NADPH necessitates the release of oxygen. The formation of ATP, on the other hand, recycles its electrons in an endless loop.

Such photosynthesizers make up the top level of the stromatolites and, as we shall see, prokaryotes capable of this double photosynthesis were the ancestors of plant chloroplasts.

Tapping into the endless source of water as a hydrogen source had a byproduct. The oxygen left behind is a waste product and escapes to the atmosphere. While much of this was absorbed by the inorganic world, eventually this byproduct appeared in appreciable quantities in the environment.

With a significant partial pressure of oxygen—and we can expect that this was higher within photosynthetic mats—a new possibility opened up. Duplicate the electron bucket brigade and disconnect one of them from the chlorophyll. Now run the electron cascade in reverse—combine glucose with oxygen to create ATP and NADPH.

Thus in a stromatolite, a layer of organisms could develop below the photosynthesizers—shaded from the light anyway—that lived off the droppings from the layer above, using their glucose and oxygen to make NADPH and ATP. This is the establishment of a simple food chain—the second layer feeding upon the upper layer. Such respiring prokaryotes were also, as we shall see, the ancestors of the mitochondria incorporated later into eukaryote cells.

WOMB EDEN 2

Organs were rapidly perfected and their forms are common to animals, fungi and plants.

It took a further billion years to generate the ingredients needed to fill in the empty QPF for the organ OS. This took place in the ocean somewhere, no doubt. Little is known about this step.

---

[1] Shapiro, R. (1985) "Origins: A Skeptic's Guide to the Creation of Life on Earth" Simon & Schuster, Inc. NY p. 88.

WOMB EDEN 3

It took a further billion years to generate the ingredients need to fill in the empty QPF for the animal OS. Again, this was in the ocean somewhere. Little is also known.

The result was the Cambrian explosion of experimental programming of the animal OS. As the programs developed in sophistication—filling in QPF that were previously empty in nature—experimentation was rife. Some were perfected, some fell by the wayside.

The evolution of the animal body plan along the branch leading to us involved fish-, amphibian-, reptile- and mammal-forms. The organ OS, as far as I can tell, has not changed. Just the programs getting more sophisticated.

It is in the nervous system where all the interesting stuff is happening, where new Operating Systems are emerging at each level.

The evolutionary pattern is the same as already established:

When all the ingredients are present—the last 'calcium' system arrives in the mix, they all jump to fill-in an empty QPF provided by nature. This QPF, as always, will obey the generalized Schrödinger equation.

The emergence of programs to run on this new OS are few at first. The most obvious source of 'seed' RNA programs is the previous level. They will run poorly on the new OS at first and their limitations will be weeded out externally.

They have little competition, however, for they are the only inhabitants of this new level of sophistication and are in a womb-like eden where everything is provided. The muticellular life that flourished just before the Cambrian explosion, but died out, was probably an almost-ran but with something critical missing in its mix. A line of thought that did not lead to anywhere interesting, so to speak.

The evolutionary pace speeds up with the emergence of programs that create a virtual reality in which programs can be tested. This is similar to the emergence of the thymus as an organ that can virtually-test lymphocytes.

Internal and external evolution now work together and rapid progress can be made in perfecting programs to run as new types of animals in the external world. Shortly, we will discuss the Adam and Eve scenario, then assume is applies on every level where sex is involved, if in less sophisticated versions..

This burst of success rapidly perfects what is possible, given the language and its inherent limitations. In the computer realm, this is like CPM and Mac OSX.

The levels of distinct mental operating systems are (at least):

The Survival or Basic Brain OS: We have inherited a basic, or fish-like sub-brain that runs all the lowest-level subprograms in a very simple RNA code-language. This involves the brain stem, the spinal column and the 'stomach' brain (the diffuse, but complex, abdominal ganglia that run our tummies). From top-down, this sub-brain is a collection of subroutines and subprograms that are called upon by higher levels and languages.

The Snake Brain OS: We have inherited a basic amphibian-reptile brain that wraps around our brain stem. Primary-color emotions—such as sex—are associated with this level of OS sophistication. The language is simple and, as it is a part of us, should be vaguely familiar. Rage and fear are others in this realm. Naturally, I am speaking of the actual experience of such emotions: Actually feeling, or worse, expressing a red-eyed, stab-and-rend, murderous Kali ferocity on a victim. Or feeling-expressing the bowel-releasing, prey-horror nausea of the living, human sacrifice.[1] There are nice things down there as well, too many to enumerate here. Talking about rage or fear, or thinking about the concepts as names, is of course, a function of the higher, human OS.

---

[1] See Temple of Doom for an uncomfortably-graphic illustration of a Kali Aztec and a living victim having his heart removed.

The Family Brain OS: We have inherited a basic mammal brain that is wrapped around the lower brains. Basic family-oriented emotions and concepts—without names—reside here.

The Tribal Brain OS: We have inherited a basic ape-hominid brain that is wrapped around the lower brains. Basic social skills involving many individuals. Names and actions, nouns and verbs, reside here as idea-of-sounds that are manipulated as a pidgin, both within the hominid VR-mind and without as sounds and gestures.

WOMB EDEN 4

Our brain, when fully functioning after about 18 years of development, generates a VR within which we, as the Main Program, run. We are capable of an internal language by which we can manipulate concepts into an infinity of possibilities. Some of them we actually do, with all the bother that physical work entails. The programs that are running in the VR, of course, think they are running on RNA in a real OS. But they are not. They are not on real RNA at all. They are divorced of any material.

When we are thinking hard, this implies, all the action is occurring in the non-material world of the VR. It is divorced from matter entirely. The neuron-firing patterns that the physiologists pick up with their scanners are just the program running that generates the VR in which we live.

The VR we live in day-to-day is generated, of course, by the physical brain. In Volume Two we will explore the possibility of the same programs running in a similar, if vastly larger, VR that is not generated by the physical brain.

It took until just 100,000 years BP to generate all the ingredients—which was the calcium factor, I wonder—needed to fill-in the empty QPF for the human mind and body.

All those different ingredients for the human OS were to be found in the different hominid races. Swimmers, climbers, upright-walking, long gestation, hairlessness, etc.—there were many ingredients that came together to jump as one to fill-in the so-far unoccupied human QPF. Simply put, miscegenation was essential to the human emergence.

When all these ingredients came together in the germ cell tetraplex—as we will shortly encounter—they jumped to the empty human QPF, then divided into male and female subforms.

The human capacity emerged from the primate recapitulation at about three when they started to create a true language out of the pidgin of their hominid forbears.

Quantum Adam & Eve

Where does a new body plan emerge? In the formation of the sperm, a tetraplex is formed.

The tetraplex jumps to a new configuration. Let us say it is the human. This is like the old Greek idea of man and woman being united at first, and then came separation.

Just so. This new tetraplex, filling in a previously unoccupied natural QPF, is the union of the male and female programs running in a virtual environment created by the recobinosomes (RNA again, no doubt.) What took so long was developing all the right ingredients in the various hominid races and then having them mix—human origins involved miscegenation.

We now imagine a multi-racial population of hominids. Through variation, these races explore the possibilities of the hominid system given the prevailing conditions. We propose during the differentiation of the hominid races there emerge morphemes that, while still being hominoid, we would recognize as one (or more) of the specialized morphemes making up a human. Such races are exploring the horizontal possibilities of morphemes such as upright posture, hairlessness, lowered voice box, enlarged brain, etc.

In the populations of the hominids, our direct pre-human ancestors, many different races will emerge by program development and innovation in the internal VR, then released for further testing in the real, external world.

Depending on the Darwinian environment, these variants will be selected for. Races will explore the possibilities of their inheritance. Some will explore hairlessness, some will explore the benefits of upright posture, some the benefits of larger brains and sophisticated vocalizations, others the benefits of opposable thumbs, etc. These can be developed in isolation but convergence is the key to the next step.

We can imagine a race that is at the center, overlapping many other races, in which all these components—a few dozen key ones suffice—mix together. We see hominids with all the human body characteristics but expressed in the hominid characteristic form.

Each of these racial adaptations, as they are called in classical science, can be expected to confer some advantages in an environment where there were many empty ecological niches to inhabit. In this sense, the environment is supportive of variation and exploration.

Speciation ✶

In the tetraplex of this endowed hominid there of four copies of the hominid somatic chromosomes, c, two copies of the hominid female master program, x, and two copies of the tweak programs that flip some things the other way and result in a male, the y.

For, it is a well-established fact that the female is the default human form. The X chromosomes do all the work of providing certain high-level images for body development and functioning. The much smaller Y chromosome does little except tweak the impact of the hard-working X chromosome. The results of this tweaking are the difference between male and female, both primary and secondary. There is nothing that a man has that is not an exaggeration or reduction in something the female has.

From this tetraplex, he makes four hominid sperm, each haploid, and two of each 'sex.'

With all the ingredients present—the calcium has arrived, so to speak—there is a probability that two somatic and two of each sex will jump to a new configuration, the configuration that is the manwoman human composite.

This now separates into two, a haploid set of human chromosomes, C, and the human master program, X; and a haploid set of human chromosomes, C, and the human tweak program, Y.

As far as I am aware, the following proposal is quite novel. The two human sets are marked with a special "speciation event" imprinting. They are condensed and packed with a "do not open yet" pattern, of heavy methylation perhaps. A massive Barre body.

The tetraplex now forms two diploid sperm, each carrying a human program in a special locker along with a normal haploid set of hominid chromosomes. This is where, science tells theology, that God's creative input arrives in the form of a previously empty QPF.

ENDOWED MALE HOMINID			FEMALE HOMINID	
tetraplex	sperm	speciation ✻	tetraplex	egg
hx	hx	~~HX~~	hx	hx
hx	hx	hx/hy	hx	3 nurse cells
hy	hy	~~HY~~	hy	
hy	hy	hy/hx	hy	

SPECIATION 1/2

The males and females of Generation 1/2 have the same father but can have different mothers.

The speciation-zygote created by their union does not unpack the specially-marked package. They are passed into the highly segregated germ cells and are probably deleted in the non-germ cell lineage, or kept as a massive, inert Barr body like the extra X in women.

"The amount of phenotypic difference between two population systems is less significant than is the presence of reproductive isolation. In fact, pairs and sets of morphologically very similar but reproductively isolated species are known in many genera of insects, flowering plants, protozoans, and other groups. Such morphologically similar species are known as sibling species. For example, the malaria mosquito of Europe … turned out on finer analysis to be a complex of six sibling species …. These sibling species, though reproductively isolated, are virtually indistinguishable …."[1]

Both male and female are normal hominids with a deep secret.

Instead of a tetraplex being created in gamete formation, there is a focus on speciation. The program comes out of the virtual world and gets to be tested in the real world. In the last act of the speciation program, the hominid aspect is deleted and only the new program-hierarchy is sent out into the world.

In the male, hominid sperm carrying a now unpacked human program are generated. This is a generation 1/2 sperm, a hominid sperm with the human program as on a CD and ready to run.

In the female, the hominid program runs the nurse cells, now only two of them, and the hominid egg is endowed with the just-released human program.

---

[1] Verne Grant, The Evolutionary Process, Columbia University Press, NY (1985), p. 202.

Depending on what "Easter Egg" was passed down the germ cells, the possible combinations are:

	MALE H, GEN. 1/2			FEM HOM, GEN. 1/2	
triplex	sperm	speciation ✱	triplex	egg	
~~HY~~	~~HY~~		~~HX~~	HX	
hx			hx	2 nurse cells	
hy			hx		

And.

	MALE H, GEN. 1/2			FEM HOM, GEN. 1/2	
triplex	sperm	speciation ✱	triplex	egg	
~~HX~~	HX		~~HY~~	HY	
hx			hx	2 nurse cells	
hy			hx		

The only real oddity predicted in this scheme of speciation is a hominid egg with a Y chromosome.

ANCESTORS

When these semi-sibling hominids mate, their hominid sperm and the hominid egg unite, as normal, to create a hominid zygote. The hominid program does a little tidying up, and then, as normal protocol dictates, turns the running of things over to the developmental program that is just starting to run.

The hominid zygote turns rapidly into a human embryo as the stuff falls into the human QPF being sequentially generated.

The Bible suggests they were not siblings, but as close as ribs are. For yes, this Generation 1/2 is nothing other than the parents of the human ancestors, Adam and Eve, at last on the scientific stage.

The basic program is: Turn cell into ball, make a hole, make another, flip them, and so on. The new programs that emerge then call on these earlier programs as subroutines. The programs pass overall control, the User effect, up the line as the higher operating systems start to kick in.

These are the first humans, born in a hominid womb, suckled at a hominid breast, raised in a thriving hominid tribe and destined to rule them all.

There is nothing to suggest that the first generation of humans number just two. A family load is quite probable. While these humans could not mate with the hominids, they could be semi-fertile with the Generation 1/2 as the Bible hints at, but their offspring would be sterile, just as with horses and donkeys.

If the first humans had not messed up, I imagine the hominids would have become human pet-servants—think dogs that do dishes—which would perhaps explain the almost universal desire for a personal servant-slave.

They did mess up—another story—and the hominids were eventually exterminated.

So, Adam and Eve did have navels, but their zygotes were created by hominid gametes and developed in a hominid womb and drank hominid food.

At four to six years old, they would have turned the pidgin hominidese into a real language, this first family of humans would have prospered and multiplied.

This sequence of internal speciation, a transition generation, and then external speciation is followed throughout the living world. I just picked on us as the example because it's such a hot topic that I hope I roundly denounced in as many media outlets as possible.

Conceptual conflict

Religion and science are offspring of the same impulse to understand what it's all about, but, like ill-matched siblings with incompatible characters, they can be at peace with each other when in separate rooms but easily brawl when sharing the same place.

Religion, at least when it's in a good mood, can be warm and supportive—giving meaning and purpose to life in the grandest of terms, giving support and encouragement, friendly and emotional. One of its character flaws, however, is that in its intermittent disputes with science, it has the most difficult time owning up when it is wrong. Just look at the retreat of religion into the petulant "He made it in six days to look as if it took ten billion years!" Perhaps this obduracy arises because it's old and venerable and science is young and brash; perhaps it's a belief that love means never having to say you're sorry.

Science, for all its cold rationality, its rejection of purpose and meaning, it nit-picking passion for collecting facts, does not have this character flaw; it has no problem—at least when all the facts are assembled—in saying to religion, "Sorry, I was wrong."

ORIGINS

One of the areas where they cannot avoid each other is origins: where did the universe come from? where did people come from? They have brawled over these two topics since science was kick-started back to life a few hundred years ago.

For a long time the bickering went something like this:

"The universe started suddenly with light!"—"Nonsense, it always existed!"

"The human race started suddenly with the first two people in one place!"—"Humbug, we came about as groups of humanoids all over the world gradually evolved into modern humans!"

Science has already gracefully conceded the first point: "Sorry, I was wrong, you were right! It did start suddenly, and light was the main event—I calculate the ratio as ten billion bits of light to each bit of matter."

Science is also coming around on the second point. It's not quite sure about it yet, but a great step in this direction appeared on page 31 of the January 1, 1987 issue of Nature, one of the most prestigious scientific journals in the world, under the heading "Mitochondrial DNA and Human Evolution." While the work was highly technical, its conclusions were starkly shocking:

"Mitochondrial DNAs from 147 people, drawn from five geographic regions, have been analyzed by restriction mapping. All of these mitochondrial DNAs stem from one woman who is postulated to have lived about 200,000 years ago...."

The authors, Rebecca L. Cann, Mark Stoneking and Allan C. Wilson, working at the University of California, Berkeley, had overcome a long and arduous course—not the least of their obstacles being the fulfillment of Nature's very strict standards—to stake their claim to a spot in the history books.

What it took to get to that point, and the reaction and rejection they received from the "old bones" paleontologists, has been documented in Michael H. Brown's The Search for Eve. Have Scientists Found the Mother of Us All? (Harper & Row, NY, 1990).

While this is not the place to get into details, we can at least lay down the general outline of what they accomplished.

MITOCHONDRIA

While most have a vague idea of what DNA is (or at least have heard about it), mitochondria probably need a little introduction.

Each of the trillions of cells that make up the body are divided up into compartments that allow incompatible processes to be kept apart. The practical wisdom of industry suggests why: a manufacturing complex—which is pretty much what a cell is—would have an overwhelming problem with quality control if duplicating computer programs onto floppy disks happened in the same quarters as burning coal to power an electric generator. Keeping such incompatible processes in separate areas makes a lot of sense

One of the great advances in the evolution of living systems occurred when a cell lineage stumbled on the great advantages of compartments and went on to become the common ancestor to all higher forms of life. The other lineages remained as simple bacteria who to this day do not have inner compartments and who, metaphorically, still duplicate their computer disks right next to the furnace.

The largest of these cell compartments is the nucleus, which is packed full of DNA. Industrially, the DNA is equivalent to hundreds of thousands of computer disks (genes) loaded with the instructions needed to program the industrial robots (proteins) that run all the myriads of processes in the industrial complex. The nucleus keeps the master disks safely stored away (chromosomes) and makes duplicates of them (messenger RNA) to send out to where they are needed in the running of the cell.

The mitochondria are usually the second largest compartment in the cell (some cells have one big one, most have lots of smaller ones). The mitochondria are the industrial equivalents of central power plants that burn fuel (glucose and fat) to generate power (ATP) for distribution to the other centers, including powering the computer-department labors of the nucleus.

All higher cells (eucaryotes) have these two compartments: the nucleus for information storage, duplication and dispersal, and the mitochondria for central power generation.

An idea that was shockingly revolutionary just a decade ago—but is now almost universally accepted—is that mitochondria are descendants of bacteria (procaryotes)—that the discovery of the advantages of keeping computer disks and coal is separate compartments involved a large simple cell (which was perhaps energetically inefficient) getting invaded by a smaller bacteria (which was energetically more efficient). While this infection was probably disruptive at first (even fatal), eventually the two learned to live together in mutual harmony—the big cell doing all the work of finding the fuel, the symbiotic bacteria, the proto-mitochondria, doing all the work of burning it up.

This insight caught on quickly because mitochondria are just like bacteria; they have their own little piece of DNA (only tens of disks-worth of information compared to the hundreds of thousands in the nucleus) and they multiply just as bacteria do: they get bigger and bigger, then split into two, with each "daughter" mitochondrion receiving its copy of the mitochondrial DNA. It is this which makes mitochondrial DNA so useful in the exploration of human lineage: its lineage is quite independent of that of the nuclear DNA.

MATRILINEAL DESCENT

The second point that makes mitochondrial DNA such a useful tool involves the way human beings are made—recall from Biology 101 that this involves the fusion of an egg cell from the mother with a sperm cell from the father.

The egg cell is huge; it has thousands of mitochondria and bulging fuel stocks all primed and ready to power the development of the new embryo. In cell terms, the egg is a big fat blimp floating lazily along, waiting for destiny to arrive.

If that destiny is not to be the flush of the menses, it will start with a single sperm piercing the egg and sparking the fabulously intricate process that ends up with a human being.

For the sperm cell, this moment of destiny does not come by waiting; the sperm has to take the gold—there is no prize for second place—in an Olympic marathon. As the run is equivalent to that from Moscow to Beijing via Mount Everest in competition with a hundred million others, the sperm can be no fat blimp; it is instead a stripped-down, sleek torpedo—just a head with its precious consignment of nuclear DNA from the father, and a powerful tail powered by massive mitochondria to push it ahead of the pack.

The single sperm that triumphs sends its head and tail to quite different destinies.

The head merges with the egg and injects the father's nuclear DNA. Inside, this combines with the mother's and is packed away into the nucleus of the cell, now a zygote, ready to provide all the information needed in the construction of a human being.

The tail of the sperm, on the other hand, exhausted from its magnificent effort, drops away, its job done, and disintegrates. The result of this sacrificial effort is that none of the father's mitochondria gets into the egg—all the mitochondria in the zygote, and the human being it eventually turns into, come from the mother.

This also makes mitochondrial DNA very useful in studying lineage: all the DNA in the mitochondria in your cells—be you male or female—came from your mother. Furthermore, your mother's mitochondrial DNA all came from her mother—your grandmother—and hers from your great-grandmother, and hers from your great-great-grandmother, etc. All the way back into deepest time.

NO SEX, THANK YOU

Yet another inducement for scientists to shift the study of human ancestry from fossilized bones to the DNA lab is that mitochondria don't indulge in sex.

Sex is the great mixer; it takes 50% of your dad's nuclear DNA and combines it with 50% of your mother's DNA to create a whole new 100% that is you. Then, in making your sex cells, it scrambles together (recombines) the contents of the dad's chromosomes with the same chromosome from the mom. That's why kids are different from their parents and their grandparents; sex keeps mixing things up in each generation.

This is the greatest thing about sex (from the lineage's point of view, at least): you get a totally different combination each generation. This blending of characters, however, is the worst thing about sex from the study-of-lineage point of view—tracing things back in time through the lineage is impossibly complicated after only a few generations.

Mitochondria don't do sex, so the copy of mitochondrial DNA which is passed on down the generations is an exact copy every time. Well, almost exact. Very, very occasionally (once in thousands of years, perhaps) a mistake is made in duplication and the DNA is changed. Most of the time, these mistakes foul things up and are quickly eliminated from the lineage. If the error is not disruptive (a neutral mutation) and happened in the formation of an egg cell, this little change can be passed on down the lineage from mother to daughter, in the matrilineal lineage.

It is these neutral changes that enable scientists to probe deep time.

Assuming that the rate of change, estimated to be 2 to 4 percent every million years, is constant—a tendentious assumption, but one that only alters the time scale—it is possible to calibrate a "molecular clock." For example, if two lineages differ by 0.3 percent, then their last common ancestor procreated roughly 100,000 years ago.

SEARCH FOR EVE ...

The Berkeley group devised a technique to isolate large quantities of mitochondrial DNA from placentas (or afterbirths, the few big chunks of human flesh that are regularly chucked away) collected from a wide variety of women representing all the races. The changes in the mitochondrial DNA were identified by snipping them into little pieces with special bacterial enzymes that are very sensitive to DNA patterns—the "restriction mapping" technique.

The assumptions they made in interpreting their results were that a particular change only happened once in history (a very reasonable assumption based on what is known) and "that the giant tree that connects all human mitochondrial DNA mutations by the fewest number of events is most likely the correct one for sorting humans into groups related through a common female ancestry," as Dr. Cann put it in her excellent overview, "The Mitochondrial Eve," in the Natural Science section of The World & I, September 1987, p. 257.

From their data they constructed a lineage that could explain the global distribution of neutral mutations. Combining this with the molecular-clock estimates and with what is known about the timing of human migrations, they concluded that the best explanation of their data was that every human being can trace their lineage back to one woman who lived in Africa about 300,000-150,000 years ago, a woman quickly dubbed "the mitochondrial Eve."

As Dr. Cann is careful to point out, their data does not prove "that all humans stem from a single female ancestor," since the mitochondrial Eve is not necessarily the very first human ancestress. There is the "Smith" phenomenon to take into account, the one that plagues telephone-directory creators—one lineage can thrive at the expense of others (though, of course, this is a patrilineal phenomenon). There could have been a group of ancestral women, all of whose matrilineal lines died out except for one, the mitochondrial Eve whose DNA got passed down to every living human being living today—it only takes one all-sons generation to stop a matrilineage dead in its tracks just as an all-daughters one will end a family name.

But the research is certainly getting close to the original ancestress. Close enough, perhaps, for science to apologize to religion for deriding the Adam and Eve concept so scathingly in the past.

In the July 1997 issue of Scientific American, the work on mitochondrial DNA had progressed far enough for the presentation of a tentative map showing how human beings spread out to populate the planet as revealed by their DNA.

... AND ADAM

What about the men?

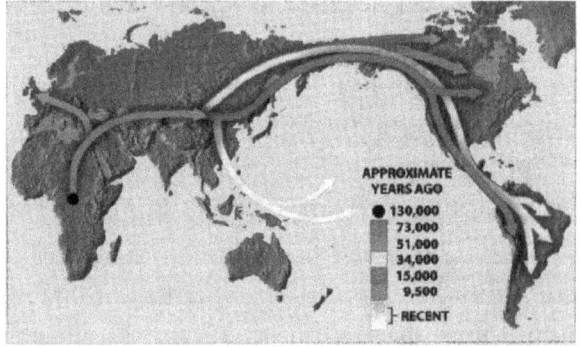

While there is no such thing as a mitochondrial Adam, there is another route. Sex determination—whether the zygote will develop into a boy or a girl—depends on what sex chromosome came from the father in his 50%: an X-chromosome will make a girl, a Y-chromosome a boy. Mothers always contribute an X chromosome: so girls are XX and boys are XY.

Boys get their Y from their dad, and he got his from his dad, and he got his from his dad, etc., etc., in a patrilineal lineage back in time.

Strangely enough, this sex chromosome doesn't get involved in sex. The X and Y that end up in a boy are so different that they don't scramble together the way the two X's do in girls. So, just like the matrilineal mitochondrial DNA in women, the Y-chromosome DNA in men is patrilineally passed on unchanged from generation to generation. Almost unchanged, that is, as it too can slowly collect neutral mutations which can be passed on. These are being studied and you can confidently expect this headline to appear one day: "Scientists find Y-chromosome Adam."

SURROGATE PARENTS

It should be noticed that science's apology is conditional: while both now agree that there was an Adam and Eve, there is still a lot of debate and disagreement as to exactly how they got there—religion still has a very difficult time with the relationship to the great apes.

Religion is going to have to unbend, sooner or later, as the mitochondrial patterns found in chimps are closely related to the patterns of mutations found in humans, which implies that the zygote that developed into Eve got its mitochondria from a chimp-like ... what?

I hesitate to use the word "mother" here as it has the implication of like to like, equal to equal. As Eve is, by definition, the first human woman, this source of mitochondria cannot be human or a "mother" in the sense of equals. But, as this female-source-of-mitochondria stood in the position of a mother to Eve, the term "hominid mother-surrogate" is appropriate.

While this does not give the definitive answer in the theological debate on, "Did Adam have a navel?" it suggests, at least, that Eve had one.

The mitochondrial linkage suggests that Eve's hominid mother-surrogate and modern-day chimps had their last common ancestor a few million years ago. Research into this is currently a hot topic of investigation.

If Eve must have had a chimp-like mother-surrogate to get her mitochondria from, you can bet that Adam must have had a father-surrogate to get his Y chromosome from.

While I have yet to see any evidence collected on this subject, bets are that the father-surrogate to Adam was also a proto-human hominid like the mother-surrogate (though, in all likelihood, they came from different lineages, since same plus same generally produces same and Adam and Eve as the first humans were, by definition, different from their parent-surrogates).

While this is speculation beyond the bounds of where experiment has reached so far, it does give hope that one day science and religion will stop their bickering about how people originated and agree that they were both partially right and both partially wrong.

Nervous OS 1.0

The lowest levels of the nervous hierarchy is quite well understood, externally. The lowest level involves the pattern of ion flows across its membrane a neuron sends down its axon, a signal down its 'output' extension that influences other cells. These patterns of electrical signals influence other cells that the axon abuts onto—which can be tens-of-thousands of other neurons in some cases. Massive parallelism is in great evidence.

The best understood aspect of how the mind works is the sensory input—how information about the environment makes it to the level of 'awareness' which, in this discussion encompasses a dog seeing a cat and racing in for the kill.

The way the senses work is that a sensory neuron responds to a 'bit' of information about the environment such as red photons, a sound frequency, a pressure differential or a chemical concentration, etc.—the senses we call sight, hearing, touch, pain, smell and taste.

Some organisms also possess more than these four, such as a sense for magnetic and electrical interactions but we humans show little evidence for such sensitivities, perhaps made up for by the possession, if spiritual experiences can be taken into account, of the ability to perceive spirits such as Jesus and Mary, ghosts both benign and malignant.[1] But enough of such speculation, we shall stick to the senses we share with all mammals.

A sensory cell is rarely quiescent. It is usually firing off a series of electrical impulses down its output axon. On stimulation by a bit of information about the environment—such as a bunch of red photons—the pattern of firing changes, a different pattern of impulses is sent of down the axon.

This can be likened to the serial connection used in computers—a modem is a good example—where the pattern of bits is sent out one at a time.

This serial pattern-change might represent a minimal piece of information—a bit, a sensory pixel, so to speak—such as 'red detected.'

These sensory pixels are analogous to the particles at the bottom rung of the hierarchy of matter. This pattern change in the serial firing of the sensory neuron is at the very bottom of the sensory hierarchy.

---

1 Examples of which are beautifully exemplified in a universally understandable tale of Scrooge and his helpers in A Christmas Carol by Dickens.

The next level in the sensory hierarchy is also quite well understood. Sets of neurons—which, in the case of the eye, are not even in the brain but in the neural nets of the retina—allow these pixels of sensory information to interact with each other with all the possibilities of interference, both constructive and destructive, so well described by complex numbers. The super-systems created by these interactions are the sensory atoms, the next level of the sensory hierarchy. In the eye, for instance, these atoms of sense are items of information such as contrast changes, color gradients, etc.

This level of representation of the environment reads: A transition from deep red to light yellow was detected.

Further up the hierarchy of programmed processing are the nets of neurons that send parallel patterns of firing along their axons to other cells involved in the next level of processing. The hierarchy of visual processing is probably the best-described of the senses, at least in the bottom-up sense of looking at things.

In the brain, neural super-neural-nets, such as the retinal columns, allow the parallel input from such as the optic nerve to interact and form higher super-systems. The internal representations of these super-systems include shapes, such as a square.

Much of the early vision information processing—in massive parallel—involves simple logic such is found in regular computers. A simple example is AND: Are two inputs the same? Yes or no.

I recall reading somewhere, that all the basic logic functions can be accomplished by arrays of NOT-AND, or NAND, that is just the 'yes or no' of AND flipped to its opposite, to 'no or yes.' The primary levels of the visual cortex do something about as simple. Many such outputs are combined into the detection of lines or patches of the same color.

The 'you' doing the seeing thinks you are "seeing the outside world." But it's a virtual reality, it's a simulation. Just like my legal copy of Windows thinks it is running a real Intel Chip, it is being 'fooled' by a simulation, it is actually running in the virtual reality generated by Virtual PC running on Mac OSX running on ... assembly code running a real Motorala chip.

You think you are "seeing reality" when you open your eyes. But it's a simulation, what is actually happening is intricately-pulsing neuron nets lighting up and fading. But the simulation sure looks real!

The visual cortex seems to be physically organized into columns of cells in which the sensory atoms integrate into more sophisticated entities. These columns of cells fire in correlated patterns when they 'perceive' things such as horizontal and vertical lines, areas of color, etc.

Sensory representations have been ascribed a process akin to the external Darwinism in classical evolutionary theory.

This so-called Neural Darwinism has gained supporters in recent years, notably Erdleman and his selection of neural representations by elimination. The law of survival in the sensory hierarchy is survival of the fittest representation. Here "fittest" implies "being a useful way of representing the reality" of the being doing the sensing—'useful' connotating the old biological mandate of survive to reproduce. A sensory image that indicates food, while the reality is a cliff is not at all useful then, in this sense.

This perspective is supported by what little is known about learning. The infant animal has its neurons in a way that can be characterized as "everyone is connected to everyone else." This plasticity is somewhat limited, of course, by the genetic constraints on the development of the brain. But there is not enough room in a trillion chromosomes—let alone the 23 of our species—to determine every one of the ways in which a quadrillion cells can connect with each other. In the totally plastic state, this number would be factorial-quadrillion which is so huge I have no idea how to calculate it.

Then there is the 'stuff' that falls into QPF in the nervous system seems to involve synchronized firing of neural nets. Are they also falling into quantum probability forms? And, if so, what might be generating the quantum probability forms for them to 'fall into'?

One possibility is the attendant, behind-the-throne glial cells that surround and embrace the well-understood neurons. As no other function except nourishment has been ascribed these mysterious "neuroglia, especially the astrocytes, oligodendroglia, and microglia" as Yahoo has it, we will not be stepping on anyone's toes.

Could RNA have a role in carrying the linear programs in the nervous system? Sure. Ten minutes with google and I came up with this:

"At learning, a sequence of events leads to a fixation of memory: information-rich modulated frequencies, field changes, transcription into messenger RNA in both neuron and glial, synthesis of proteins in the neuron, give a biochemical differentiation of the neuron-glial unit in millions, a readiness to respond on a common type of stimulus.

"At retrieval, it is the simultaneous occurrence of the three variables: electrical patterns, the transfer of RNA from glial to neurons, and the presence of the unique proteins in the neuron, which decide whether the individual neuron will respond or not."[1]

"In neurons, localized RNAs have been identified in dendrites and axons; however, RNA transport in axons remains poorly understood... It is concluded that the specific delivery of RNA to spatially defined axonal target sites is a two-step process that requires the sequential participation of microtubules for long-range axial transport and of actin filaments for local radial transfer and focal accumulation in cortical domains."[2]

Neural OS
RNA linear program runs
Generates Quantum Probability Form
Stuff falls into collapsed probability form **Organ OS**
Cell OS
BASIC OS

To My Mind

We will equate this with the emergence of the capacity to invent a grammar. This is a rather specialized aspect of language that adults do not have, inasmuch as they have "lost" it by the teenage years. Lost is probably not the correct expression, however. Rather, higher structures have come to depend on the constancy of grammar rather than its infant mutability. If this is correct, the "ontology recapitulates phylogeny" perspective suggests that this childhood stage is an echo of the Origin of Man.

That children temporarily have the faculty to invent grammar while adults do not became apparent when linguists investigated the origins of new languages in historical societies. The surprise was that only children are involved in the origin of real languages. This faculty is usually hidden since most children grow up immersed in the language of their parents: they learn that with remarkable facility and do not need to invent a new language with a new grammatical structure.

The rare exceptions to this—where children were not immersed in the grammar of their parent's culture—are where languages have been invented in recent times.

One example of language invention involved deaf children in a large institution in Central America. They were not immersed in the grammar of the adults, and so invented their own. They transformed the primitive pidgin signings of their few adult teachers into a true language with a fully-fledged sophisticated grammar.

As far as regular vocal language is concerned, there are many examples of true language invention in history when children developed in a culture in which a pidgin is spoken. Adults invent pidgins, they do not invent languages. A pidgin is not a true language in that it does not have a grammatical structure that can express any but the simplest noun-verb combos.

Pidgins have been invented by adults many times in history; they are quite common. When adults speaking many languages are forced to live together—as in port cities or slavery situations—they spontaneously develop a pidgin that allows for basic communication and economic interaction to occur.

---

1 Hyden, H., "The question of a molecular basis for the memory trace." In Pribram, K. H., & Broadbent, D. E. (eds.) Biology of Memory. New York: Academic Press, 1970, p. 116.

2 Ilham A. Muslimov, Margaret Titmus, Edward Koenig, and Henri Tiedge, "Transport of Neuronal RNA in Axons," The Journal of Neuroscience, June 1, 2002, 22(11): 4293-4301.

A pidgin can convey basic information about things and actions; but not much more. In a pidgin, "John kill Jim," "kill John Jim," "john, Jim kill," etc. all associate a death with these two individuals; but it can convey no more. It is not possible to pass on a full description or understanding.

Nevertheless, a pidgin can be remarkably effective in allowing for basic social interchange in a polyglot population of adults. A pidgin is not capable of describing exactly who did what to whom; there is no grammar; there is no subject-verb-object structure to slot the words into. We suggest that pidgin was the highest linguistic ability of our ancestral hominids. They had sounds to represent objects and actions but no way of stringing them together into linear strings with a grammar structure.

Children developing in a pidgin-speaking environment are not exposed to a grammatical structure, they develop in a grammar-less world. They first learn all the sounds used around them; then they pick up all the pidgin words in use around them; and then they do the unexpected, they effortlessly invent a grammar; they organize the pidgin into a true language. They invent a true grammar and transform their parent's pidgin into a true language, a Creole. The Creole is the simplest type of true language—it matures by adding new words and speakers into a "regular" language. Adults do not have this innate capacity to improvise a grammar on the fly or effortlessly learn the language. Children exposed to a grammatical language use the innovative capacity to effortless language acquisition. In either role, the faculty is lost in later years. As adults, we can pick up new languages but only by strenuous effort and we can invent languages but it takes university-honed skills as a linguist to do it.

Grammar is like putting our thoughts in order. We think in language. I am sure that the deaf-language innovators thing in terms of sins.

The scientific worldview we will be constructing on the internal cause-of-probability of quantum physics has a certain resonance to a philosophical structure created by Karl Popper. In his classification, the objective reality studied by scientists—atoms, planets, cells, galaxies, brains, etc., etc.—corresponds to World One in his profound philosophical dissection of reality into three realms.[1]

Popper's World Three is the realm of the mind, what we have going on inside us. For example, the concepts and theories of science belong to this realm. This World is what goes on inside each person, the thoughts, theories, concepts, plans, emotions, passions, etc.

World Two is where World One and Three intersect as ideas are expressed in life and culture and, occasionally, in science. Expressions of scientific thoughts, plans and passions in the form of books, educational institutions, cyclotrons, conferences, etc., all belong in Popper's World Two. World Two is the expression of human thoughts, ideas, plans and passions in all that we see about us—buildings, washing machines, concerts, newspapers, dollar bills, interstates, etc.

One general way of interpreting this philosophical perspective is that human artifacts—which in terms of classical random chance-and-accident are highly unlikely aggregates of atoms—can only be comprehended if an internal influence is included in the discussion. The aspect of culture we call science, for example, can be thought of as scientific thoughts influencing what happens to scientific materials.

While I don't think Popper's concepts embraced the notion of probability being the fundamental link between his three worlds, the similarity between the two is apparent. If the similarity holds, modern science leads us to expect that ideas in the mind are linked to probability. Ideas in the mind are probability forms that manipulate matter: the idea for Mona Lisa provided the probability form for the oil paint to "fall into" in the painting process.

Does the human speech module, the human speech program involve programs encoded on RNA? I think so.

---

[1] Sir Karl Popper, Objective Knowledge, Oxford University Press, Revised Edition, 1979. First published: Oxford University Press, 1972.

First, the program gets run in the virtual reality generated by the Main Program. To experience this directly, think this thought silently 'inside:

"What I think of as 'reading this sentence in my mind' is actually RNA programs running in a virtual OS environment. The 'I' doing all this is actually a Master Program."

Most of us have a limited success reprogramming this Master Program, but it is usually difficult.

Just like the testing program running in the thymus, you can release these programs from the virtual reality to the real. To experience this, read this sentence silently until instructed to speak:

"What I think of as 'reading this sentence aloud' is actually RNA programs running in a virtual OS environment that I am releasing to the real OS where they are running and I found myself speaking this sentence fragment, 'releasing to the real OS to run as the speech fragment....'."

Did you start speaking on the first thought 'releasing?' If so, did you catch the "release" command you gave. It's kind of hard to do at first as we are so used to either reading silently or reading aloud so the switch in midstream is unpracticed.

Try it a few times. Try whispering it.

The stored programs are probably in my cerebellum which seems to have a syntax checker as only well-formed programs are allowed into storage—'learning' or 'getting it.' Reading is a major program and takes a good while to assemble.

Learning, the, is the putting together of a program that runs. Once it runs properly, we have learnt it and it is stored in the "DNA" cerebellum for retrieval when called on.

Are quantum probability forms involved in the human mind, I wonder? A simple experience of mine leads me to think so:

OSSINING LONGING

For over ten years, I commuted from Ossining to Manhattan on the MetroNorth train. Almost every day I was assaulted as I waited for my morning train with the strident computer-announcer insisting:

A t t e n t i o n !   A t t e n t i o n   a t . . . !
O s s i n i n g !

For 10 years, every day, this voice resonated inside my skull as I fretted, "What now!" I have now, as of this writing, lived in Mount Kisco for fourteen months. I still catch the MetroNorth. Almost every morning the same voice intones:

A t t e n t i o n !   A t t e n t i o n   a t . . . !

And what is strange is that there is an "Ossining" shaped hole there in my head that, for a moment, is very perplexed when the Voice says:

. . . M O U N T   K I S C O !

For it does not fit. It's empty, a pixel of frustration.

It happens to me every day and there is nothing I can do to stop that little glitch of surprise. Could such an expectation be an empty probability form? Is a program running in the front of my brain (cerebrum) and generating an "Ossining-shaped" empty probability hole with a 'desire' to be filled and not empty? A Pavloved-dog probably felt the same way when that darn bell rang but no food appeared.[1]

Now, if a part of my mind is made of probability forms, perhaps a lot more of it is; perhaps the whole shebang.

---

[1] Kerry Pobanz, my philosophical advisor, informs me that, "This is reminiscent of the treatment of habit, habituality and novelty by Rupert Sheldrake, which, in turn, is rooted in the Process Philosophy of, especially, the foremost American philosopher, Charles S. Pierce and the thought of Alfred N. Whitehead. Thanks, Kerry.

So, the QPF of the aminoacids are a distant cousin of the QPF for "Ossining" in my brain. Perhaps my anthropomorphic translation of quantum math into natural English has some justification after all.

In one of those delightful moments of synchronicity, as I was writing this a commentator on a WNYC spoke on studies of how familiar music runs in the head. A specialized part of the front brain (the auditory cortex) was highly active (running a program?) and remained active even when the catchy tune (The Pink Panther riff was the example) suddenly stopped midway. Something was there with no sound in it. When an unfamiliar tune was played, however, the activity stopped the moment the tune stopped.

BACTERIAL FEELINGS

Actually this kind of feeling is not as sophisticated as we might imagine.

Consider a bacterium. As established, it is an internal composite of quintillions of QPF generated mainly by proteins. Of these, a fraction generates QPF for glucose, say. In a healthy, well fed bacterium, all of these trillions will be filled; they will be 'satisfied' in an internal way connected to the Path of Least Action.

But in a difficult environment, perhaps only 5% of these glucose QPF will be filled.

What about the 95% of the quantum probability forms that are empty? All the ones with nothing in them. Does this void amount to anything?

Now, in the classical view, the concept of a bunch of nothing amounting to much is quite ludicrous.

But, we know, from our weird execution, that a bunch of nothing can indeed amount to something very significant, like bulletproof vests made of nothing but a void.

So, what do these trillions upon trillions of empty QPF amount to as they clamor to be filled? Just like a simple aminoacid in a chain, just on a larger scale.

Could not this unhappy bunch be a primitive kind of feeling of hunger? The part of the overall composite that is 'not happy' with the way things are, is similar, indeed, to our own basic instincts, except in depth and scale, that I withdraw my earlier apology for using anthropomorphic analogies to describe the longings of aminoacids.

BACTERIAL AUTONOMY

We earlier established that an electron is ascribed a simple autonomy in quantum physics. It is known in the labs that even God cannot know where the electron will land in a slit experiment.

We linked this autonomy to why it is impossible for a computer to create random numbers. The best computers can do is generate pseudorandom ones—"random numbers generated by a definite, nonrandom computational process" according to Yahoo. Clearly pseudorandom is not really random at all. Yet, you can reel off a string off random digits no difficulty.

Electrons, when faced with 'competing' probabilities—such as 50% go this a-way, 50% go that a-way—have a true autonomy.

We expect no less of bacteria. When faced with competing probabilities they also can be expected to have a true autonomy. The mind is mysterious, even in bacteria.

THE HUMAN OS

The mind is definitely to be found in layers, reptile, mammal, etc. But, simply put, we can add just one more level to the edifice.

Neural OS
RNA linear program runs
Generates Quantum Probability Form
Stuff falls into collapsed probability form **Organ OS**
Cell OS
BASIC OS

For now we can think of these mental atoms in terms of what science has already established—or rather not established—about the "binding problem." We can illustrate this with the visual system. The challenge is that we know how the base of the visual system detects all sorts of "primitive" things about the world—edges, lines, and transitions in color, etc.—in many different areas of the brain. The challenge is to figure out even a good suggestion as to how the visual system integrates all these primitives together to "see" a distinct object such as a tree rather than just a lot of lines and colors.

We will look at this in terms of filling in of empty wavefunctions—coded images corresponding to things in the environment. For the human capacity, we will further explore the idea that these images become "things in themselves"—they become the atoms at the foundations of some higher intelligence. We will equate this with the phenomenon of "naming"—being able to think of a "tree" without actually looking at one. The image has a reality that is quite independent of there being trees to look at.

LEARNING

It takes a human being about three years to assemble an "abstraction module" that can recognize an abstraction in the following story of a five-year-old's birthday party.

A short story:

Richard was five. Aunt Marie and Uncle Willow were there, Aunt Betty (Peter, her husband had recently passed, so there was a little gloom about), and dad's spinster sister, Aunt Francis, and lots of friends.

When his mom asked him later about the party Richard, being a smarty pants, declared "there were three aunts here for three hours and I was so nice to them. But all I got was three shillings, and three kisses too. Yuck!"

We get so used to such counting skills, like riding a bike, that seems automatic and hence, common sense. For the little boy recognized that three aunts, three hours, three kisses and three pennies—which otherwise have nothing else in common—do have something in common: the abstraction, three.

Such a familiar skill is difficult to view as special, but animals do not have an abstraction module, so see nothing in common between three prey and three predators.

It seems that the front of your brain—the cerebrum—is where you struggle to assemble new and sophisticated programs (concepts) while all your automatic ones, the ones that actually require effort to become conscious of, are in the back, the cerebellum. This is where the 'riding a bike' that "you never forget" is stored and can be called upon effortlessly as an RNA linear program is run.

All the learning takes place in the front, which is constantly trying to get the cerebellum to store it away. The problem is, the cerebellum is where programming rules are strictly applied and it only "accepts" what mathematicians call well-structured constructs. It is a syntax-nut, nit-picking module that will only run well-structured programs. This means, in essence, it will only accept programs that have already been proved to work in the VR up front.

The front, in the cerebrum, is where "you" speak English silently. The backroom cerebellum is where math-speak is spoken. The front can think $1 + 1 = 4$. The back will not accept it, it will reject the malformed program and wait for you to come up with $1 + 1 = 2$, a well-formed statement that is eagerly admitted into the math-speak program store. Somewhat like DNA.

There is nothing quite like the feeling of getting it, especially if you have been dumb enough not to get it while all the girls have already. But once you get it, you know from experience, you never forget because it's stored safely in back.

Sleep

My Mac has a lot of housekeeping chores it has do to do—update the clock, flush memory to disc, adjust the virtual memory, etc. It has lots of such programs to run. Being a 'threaded' CPU, each gets a few microseconds every second to run and do its thing. The chores are important; if they are not done, the big shot programs that get milliseconds will quickly grind to a halt.

Apparently, such rapid interpolation, such intimate mixing of necessary chores while running the Main Program is not possible when super-massive parallel processing is involved. For all animals with a brain have to sleep.

It looks like the chores have to be done at night; the Main Program—that's you, dear reader—is shut down and all the housekeeping programs come out of storage on neuralRNA and get to run in waves of shifts, such as REM sleep.

That 'refreshed' feeling that is so desirable when the alarm rouses the Main Program back into action is no ephemera. QPF that should be empty are empty. QPF that should not be are not. That moving finger Main Program that is "I am" starts to Run and generate probabilities that the day will be a good one as the external me interacts with others on another day on planet Earth.

Why is sleep so important? I would not suggest going without sleep for many days. For soon, no matter what kind of Main Program you are, you will start to experience "system crashes," things like not knowing who you are or why you are in this room that is unnervingly psychedelic. (Perhaps LSD triggers the running of a housekeeping program that, at least for me once, makes the Main Program believe in magic, and speaking to trees.)

This is a screen grab of my activity monitor as I type. The only two programs I am aware of using are Word and Finder. What are ATS and PBS doing in there with all those megabytes to sprawl around in?

If I were to somehow disable all these OS X chore programs, both Word and Finder would almost instantly on my scale of things grind to a miserable death, trust me.

Just so, without sleep, your Main Program consciousness will falter and crash badly, sooner or later. Getting a good night's sleep is as good as control-alt-delete is as in resurrecting a crashed Windows.

The brain's neural operating system keeps on working throughout the night—it's almost as active as when awake, just a very different kind of activity. But it is busy running all the many

Process Name	% CPU	Threads	Real Memory	Virtual Memory
Word	11.40	2	40.36 MB	241.12 MB
WindowServer	2.40	2	39.37 MB	211.46 MB
Finder	0.00	2	16.78 MB	164.95 MB
MenuStrip	0.40	5	15.16 MB	158.57 MB
ATSServer	0.00	2	14.84 MB	87.43 MB
Activity Monitor	2.90	2	10.85 MB	161.69 MB
Grab	5.90	3	5.89 MB	149.58 MB
loginwindow	0.00	5	5.27 MB	124.21 MB
SystemUIServer	0.00	1	5.02 MB	148.00 MB
Database Daemon	0.00	1	4.43 MB	138.98 MB
Mirror Agent	0.00	3	3.43 MB	144.57 MB
Dock	0.00	2	3.42 MB	139.07 MB
iCalAlarmScheduler	0.00	1	2.81 MB	100.33 MB
pbs	0.00	2	2.23 MB	44.41 MB
iTunes Helper	0.00	1	1.13 MB	91.36 MB

housekeeping chores; it is not running "you" anymore. The Main Program has been "written to disk" on virtual RAM, ready to commence running again in the cleaned-up real RAM in the morning.

In the old OS, a program crashing also crashed the operating system (which is mostly virtual) and I would have to pull the plug, count to three, and then restart it. Horribly time consuming.

Now, however, with OS X, a Force Quit program is (usually) available under the apple (thank you Steve, great job. Me and my 400 apples are grateful.)

This simple program must issue some very low-level instruction to OS X such as "terminate that program and wipe its data structures in its RAM partition." No matter that Word has totally frozen, and the hypnotic wheel is spinning endlessly. At most, I will have lost a paragraph, sometimes just a word, for I save compulsively from bitter experience.

I just calmly hit the right keys and this little window is generated. A click and all is new.

Going to sleep is just like that. Some low level timing program notes that it is time, and safe, to sleep. It runs the "Force Quit" program and clicks on Main Program. The Main Program instantly stops, the contents are written to disk on RNA, and you go to sleep.

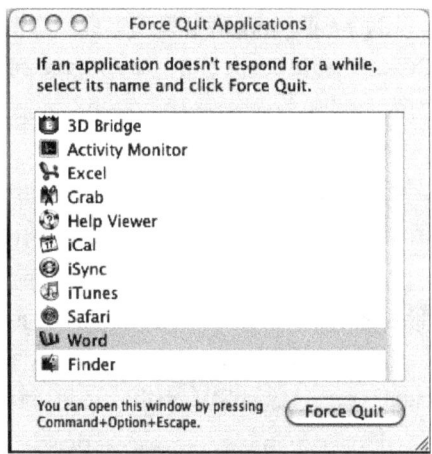

As is well known, we cannot will ourselves to sleep. Going to sleep happens to us not by us. For this reason, it is not possible to actually experience "going to sleep," we just abruptly stop. There is no "you" there to notice what going to sleep feels like. The housekeeping programs get to run and start cleaning up and moving stuff around.

One of the side effects of these housekeeping programs is dreaming. This is probably bits of the day and other debris running momentarily in some module. The characteristics of this type of dreaming is that we are not in control, and in fact it's more like fragments of "I Am" running with a limited autonomy that is not "I". Our bodies move a little; dogs scratch and sniff. The memory of such dreams rapidly fades within minutes. They can reveal you something about the fragment of yourself that was running, but little else.

There is another type of Dream, however, that is qualitatively and quantitatively quite different from 'animal' dreaming. This type of dream is very vivid, we are our whole selves, and we can speak. It is as real as the real world. In fact, on waking from such a vivid and real Dream it is the bedroom, the everyday world that seems unreal, dull and monochrome. The Dream is so real we expect it, not the bedroom, to continue.

I probably do not have to remind you that such Dreams can be nice, or they can be not nice. In fact, a real nightmare can be totally unsettling. When I was young, I had the same Nightmare many times: a hideous witch screaming after a terrified, running me. She never caught me. The dream was so utterly and vividly dreadfully real that I was shaken and miserable for weeks after each occurrence.

Then there are the glorious Dreams, where everything is perfectly and delightfully intense and wonderful. I have flown in such Dreams and once or twice flown in day-Dreams. If you have not had such a Dream, the best art that captures it for me is Disney's Peter Pan where they soar above Victorian London. The 'real' thing is much, much better. And it's so easy—the "up body" impulse is as natural as the "raise arm" one is.

'I Am', Fermat & Hilbert

The leap from the hominid to the human mind could be as simple as a programming language which goes from manipulating just real numbers to being able to manipulate complex numbers.

The first thing I do when I get a new calculator program is ask it to display the square root of minus one. Almost always, they say something like "not a number" and I sigh with frustration.

For it is safe to say that all the math used in science and technology involves complex numbers at one stage or another. So, I find this restriction of simple calculators to the real numbers hard to understand.

I am quite sure that it will be found that the human VR mind and 'I Am' Main Program all manipulate complex numbers, not just real ones.

Self Creation

Incidentally, you might be wondering where this "I Am" program that is you came from. It has been assembled, code by code, by yourself as you have lived your life. The human program is self-developing.

Naturally, much of the first layers of the "I Am" master program, the ultimate User, were laid down by your parents in the very early years. But as your sense of "I Am" my own person emerged, more and more it was you who assembled the code by the choices you made and things that happened to you. Naturally, the culture larger than the immediate family also plays an increasingly important role.

By 16 or so, it is a safe bet to say that you were making all the really important decisions by yourself, and a few 'best' friends and cultural role models (for both good and ill.).

So the current "I Am" program was added to by what yesterday's "I Am" experienced, which was added to by... etc. This is why we can be so complicated inside. We are a labyrinth of a program, layers upon layers upon layers, forgotten or forbidden-to-run places, etc. And, let's face it, some of the subprogram routines have bugs in them, some so awful they cause you to "crash" when the run in your mind like an old tape replaying yet again.

Unlike a simple computer, a massively parallel one such as the brain, can run two, or more, Main Programs at the same time. Perhaps the comment, "I have a divided mind..." can be taken literally; two slightly different "I Am" versions running at the same time. This is never a comfortable situation, and it can get really bad for some unfortunates.

Externally, I might be what I eat. But internally, in the realm of QPF generation where it really counts, the Main Program that is "I" was assembled by what I did. So the "I Am" program that runs everyday is a composite of all my life experience; some active, some quiescent but (waiting to spring to life, sometimes at very inconvenient moments).

This is the "I Am" program that is 'written to disk' every night when I fall asleep. Everything that is me is written onto an array of linear RNA programs.

My Mac OS calls this memory dump file that it creates when going to sleep, the VMfile. If a lot of programs were running when I closed the lid, this file can be many gigabytes in size on my hard drive.

So, that is where "I Am" is when I am asleep, stored in an organic VMfile in the form of RNA arrays of arrays. It will involve a truly-huge amount of linear information. This must involve a truly immense amount of information: whatever-bytes upon whatever-bytes of linear information all copied out of RAM and onto the RNA 'disk.' Each glial cell probably has just one "I Am"-RNA. There are a zillion of these, so a zillion RNA are involved; small if 'empty,' as big as necessary to store some complex experience.

The storage capacity of even milligrams of RNA, however, is really, really astoundingly huge, not one of our G numbers perhaps, but getting there: it's on the order of 4 to the power of Avogadro's Number, $4^{10^{23}}$. This expression is so huge that, just to reduce it to two stories high, 10^N, to write out N as 1,000, 000... I would fill the visible universe with them, even if I made each zero the size of a virus.

So, a milligram or so of RNA is more than adequate to store all that is the "I Am" program that is you or me. Everything. Nothing is edited out or deleted (a selective Erase module would be nice for deleting the bad stuff—but the 'moving finger' has no 'backspace-erase' function, unfortunately.

RNA Arrays

There is both good news and bad in this perspective for push-the-envelope brain scientists:

Good News: An entire human personality can be stored onto, and retrieved from, on a few milligrams of RNA[1]. Subprograms, like riding a bike or speaking Korean as a second language, are also stored on even smaller amounts of RNA.

Bad News: Taking RNA as a pill is unlikely to work (at least with currently-feasible technologies). For the RNA molecules that are doing the storing are arranged in a precise array. The precise array corresponding to the positions, or 'addresses' of the glial cells. For the brain is not a jumble of cell, it has an intricate, if highly-repetitive, array of cells in layers and lattices.

---

1 See Gateway by F. Pohl for a SciFi prophecy on this matter.

Location is everything, as is well known. The hard drive on this Mac is divided into sectors that are sequentially numbered. The OS only deals with sectors, reading and writing to them as needed by calling up the position, the address, of the sector.

From the point-of-view of Mac OSX, a sector can hold a wide variety of things such as a bit of: data, a command, a word, a picture, a sound, etc.

From the point-of-view of the sector, it's all the same: a pattern of tiny N and S magnetic poles (or tiny pits and no-pits on a CD; really tiny bits and no-bits on a DVD, etc.)

So two things are important when you go to sleep—when the "I Am" program running in the VR is written to disk for the night's safe storage.

The subprogram written on the glial cell iAm-RNA.

The position of the glial cell in the multi-dimensional layered structure of the brain; its address.

Escher captured the idea:

A simple way to describe this is to use Hilbert multi-dimensional matrices. A simple 2-D, 2x2 matrix looks like:

$$\left\{ \begin{array}{cc} +A & +Bi \\ -Bi & -D \end{array} \right\}$$

The manipulation of matrices, with thousands of rows and columns, is a well-understood aspect of math and is used extensively in much of science and information technology.

High-end calculators, such as Mathamatica, can not only deal with complex numbers, they are proficient in matrix algebra as well.

A simple extension of this allows for a complete description of the RNA-glial cell array that stores the "I Am" as one goes to sleep.

First, the 2-D matrix is generalized to have an unlimited number of real dimensions, each at right angles to all the others. A real hypercube, a hypermatrix. At each location, there is a value. For the brain, this is the iAm-RNA program chunk stored in a particular glial cell during the night's rest.

All of these axis are real. This is like putting a line at a right-angle to another in 2-D space, then another at a right angle to both in 3-D space, then another at a right angle to them all in 4-D, etc.

The next step is to put, at a right angle to each real axis an imaginary axis. This is called a Hilbert space of complex dimensions.

So a 3-D Hilbert space actually has six axis, all at right angles to all the others—three that are real and three that are imaginary. Combinations of the two can be considered as a complex axis. Tensors are good at this kind of stuff.

Just to illustrate how useful such multidimensional Hilbert spaces are we shall take a short break to use them to solve a little—well, bigger than a margin—problem.

Fermat's Theorem

Fermat's conjecture involves this relationship between numbers:

$x^p + y^p = z^p$

Where x, y, z and p are positive real integers greater than zero, and p is a prime.

The hypothesis is that while there are an infinity of solutions for the solitary, <u>even</u> prime number, $p = 2$, there are absolutely no solutions for the denumerable infinity of <u>odd</u> primes starting $p = 3$, 5, 7, 11, etc.

The first point is that a real integer raised to a real integer is always a real integer, and this holds true all the way out to this side of infinity. The second point is that:

If:

The number x measures a line in 1D-space

The number x^2 measures an area in 2D-space

The number x^3 measures a volume in 3D-space

The number x^4 measures a hyper-volume in 4D-space

Where each x-axis (or y and z) in the x-hypercube is at right angles to all the other x's in the cube. In the unit cube in n dimensions, there will be N points with a coordinate 1 and just one unique zero point. In an infinite-dimensional, the will be on infinite cloud of 1 around the one unique zero point. In an infinite-D Hilbert space there will also be an infinite could of points at I, but still one unique zero point.

Then:

The number x^p measures a hyper-volume in pD-space. Same for y and z.

So, Fermat's relationship is actually about adding hyper-volumes and then equating them to another hyper-volume. This is just a generalization of the way Pythagoras equates the area on the hypotenuse and the sum of the areas on the other two sides.

Translating all this into English, Fermat's equation states that the hyper-volume of the z-cube is measured by a real integer that is equal to the sum of the hyper-volumes of the x- and the y-cubes when added.

In order for the x and y cubes to simply add together as they do, they must be independent and distinct hyper-volumes with no overlap. For this to hold, every x-axis must be at 90° to every y-axis while also being at right angles to all the other x's. So, our hyper-volume inhabits a <u>complex</u> p-dimensional hyperspace. I believe the technical term for such is a Hilbert Space.

HILBERT SPACE

In Hilbert Space, the relationship between x and y is that of a real and an imaginary axis. This orthogonal requirement is satisfied in two simple ways. For every x there is a yi or, for every y there is an xi. It makes no difference, so we will use the first.

We can add this to our list above:

The number $x^p + (iy)^p$ measures a hypercomplex-volume in pD-Hilbert space. (A pHD-space?)

We need just a few aspects of Hilbert space to make our point. All the axis, both real and imaginary, touch at just one point, exactly zero. All extensions in this space start at this common point (you cannot have an extension along any axis starting at $-1/2$ and ending at $+1/2$.

Most mathematicians keep track of which axis is which by using a cumbersome convention called the right hand rule (or is it the left thumb, I forget).

Much simpler is to accept the well-defined concepts of complex areas, complex volumes and complex hyper-volumes into the descriptive arsenal. I have already accepted complex 1-D lines, as complex number little arrows, so why stop there.

Consider the following four squares, each with sides 1 unit in length, and the following questions about it.

What is the area of the four squares if:

x is a real axis and y is a real axis?

x is an imaginary axis and y is an imaginary axis?

x is a real axis and y is an imaginary axis? Or the converse.

The answers are, with x going first:

Both x and y are real: Doing the usual math we have:

+1 x +1 = +1

+1 x –1 = –1

–1 x –1 = +1

–1 x +1 = –1

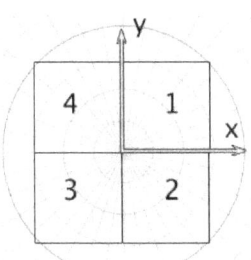

Yes, in Hilbert space, areas can be negative. As most quantum physicists find that a Hilbert space, not a real space, describes the way the world really works, this is no small matter

Both x and y are imaginary: Doing the familiar math we have.

+i x +i = –1

+i x –i = +1

–i x –i = –1

–i x +i = +1

The areas extended by two imaginary axis at right angles are also real, just with the signs reversed.

Real x, imaginary y—the symmetry showing why switching roles would have no effect on the outcome.

+1 x +i = +i

+1 x –i = –i

–1 x –i = +i

–1 x +i = –i

So the area extended by multiplying a real axis by an imaginary axis (technically, the Cartesian Product) gives us imaginary areas. What else?

As i is also known as the rotation operator, an imaginary area is at right angles to real areas.

This combination of a real and imaginary axis is the complex plain, so the regular complex plain is actually a 1-D Hilbert space.

HYPER-VOLUMES

There is a subtle difference between a regular space and a Hilbert space that very is important for our discussion. This is because 50% of the Hilbert is composed of imaginary axles.

Construct the infinite real unit 'hypercube' in the following manner with each side of positive unit length: a 1-D line, a 2-D square, a 3-D cube, a 4-D hypercube… a countable-infinity-D hypercube.

The sequence of hyper-volume is:

1 1 x 1 1 x 1 x 1 1 x 1 x 1 x 1 …

Which is 1, 1, 1, …

Now construct the same infinite hypercube in an infinite Hilbert space, but this time using all imaginary axes.

The sequence of hyper-volume is now:

i i x i i x i x i i x i x i x i …

Which is i, −1, −i, +1, +i, −1 …

The hyper-volume of the imaginary nD-unit cube is:

Imaginary for all n-cubes with an odd number of edges out to infinity (and beyond).

Real for all n-cubes with an even number of edges.

This essential distinction between volumes with an odd or even number of edges is not to be found in a regular space, only in a Hilbert space.

The hyper-volume of the imaginary nD-unit cube is rotating counterclockwise, the positive direction with a period of four through the sequence of volumes that are:

+IM, −RE, −IM, +RE, +IM, −RE, −IM, +RE,
+IM, −RE, −IM, +RE, +IM, −RE, −IM, +RE,

This is a very interesting cycle that seems to reflect a basic fact of life: the three dimensions of space and the one dimension of time.

Euclid used real numbers to describe spatial separation, length, area, volume, etc. Pythagoras codified the relationship between axis components and 'straight line' separation which, generalized for any number of orthogonal real dimensions, is:

$s^2 = x^2 + y^2 + z^2 \ldots$

Einstein added time to the familiar three of space, but he assigned it an imaginary axis to extend along. In the equations of general relativity, time appears as an extension along an imaginary fourth dimension. So time always appears as 'ti.'

The connection between components and separation in Einstein's unified spacetime is now:

$s^2 = x^2 + y^2 + z^2 + (ti)^2$
$= x^2 + y^2 + z^2 − t^2$

Plus-one and minus-one are distinctly different and easily distinguished. On the other hand, plus-i and minus-i are so identical that they can be switched with impunity. The distinction is not significant, it is just a convention. While there are two distinct real units, +1 and −1, there is only really a single imaginary unit, i.

The complex conjugate of a number is the same number, just with all the plus and minus signs to "i" flipped. This 'reflection' of a complex number in the real line is the one that appears in the collapse of the wavefunction when a complex numbers transforms into a real number.

Back to the four-cycle of hyper-volumes in a purely imaginary space.

Start with "i"—or NOT real—and call it time, an imaginary extension.

Rotate this unit imaginary extension by 90°—multiply it by i—and you have swept out a negative-real area in Hilbert space. Call this new orthogonal axis the 1st spatial dimension.

Rotate this by another 90°—multiply it by i—and you have swept out a negative-imaginary volume in Hilbert space. Call this new orthogonal axis the 2nd spatial dimension.

Rotate this by another I and you have swept out a positive real volume in Hilbert space. Call this new orthogonal axis the 3rd spatial dimension.

As positive-real is about as real as it gets. We seem to live in a reality that is just filled with things that can be described with real, positive numbers. And all the complexity of every wavefunction collapses, in the end, to a real, positive number.

We live in a veritable bubble of a real, positive universe. With just these four dimensions providing us with a "just right" positively real environment to inhabit.

From above, we conclude that the recipe for creating our universe—or at least the spacetime 'stage' on which all the interesting stuff can happen—might have been as simple as:

Take one i from Hilbert space.

Cube it.

Voila, a positively real spacetime.

In Hilbert space, the relation between components and separation/distance is:

$$s^2 = (ti)^2 - x^2 - (yi)^2 + z^2$$
$$= -t^2 - x^2 + y^2 + z^2$$

So why did our universe extend to infinity into four dimensions, and only four, starting at the Big Bang? Why did just four do it after the Big Bang? The others clearly didn't, and those other dimensions remain un-inflated and Planck-sized to this very day—and there seem to be at least eight of them. These are curled-up, multidimensional strings and branes extending a tiny distance from every point in regular spacetime. (They are hardly noticeable but the Higgs finds them a fine place to deconstruct in, as we shall discuss in Volume Two.)

The answer probably goes something like this:

Eight dimensions would be positively real, but would be too confusing for simple folk, two directions in time would be particularly so. Twelve dimensions would also be real, but even worse with three time-like extensions (nightmares sometimes exhibit such a poly-dimensional time and space experience; and it's not particularly pleasant.)

One, two or three dimensions would not be positively real. Thus, four is the only one.

FERMAT IN COMPLEX HYPERSPACE

If Fermat's is true in a real space, it will certainly be true in Hilbert space. If false in Hilbert space, it is most certainly false in real space.

With Fermat, we are actually adding two complex volumes and equating them another volume

All these volumes in Hilbert space have a common point at zero and they have to be non-intersecting. We can only accomplish this in arbitrarily-large dimensional hyperspace if one volume, say x, uses all real axes while the other uses all imaginary axes.

So, Hilbert Space, Fermat's relation is actually:

$$x^p + (iy)^p = z^p$$

where z^p is a hyper-volume measured by a real integer. For the solitary, <u>even</u> p, this reduces to the expectation that:

$$x^2 - y^2$$

has to be a real integer. As this is a trivial expectation, we pass it by without comment.

For all <u>odd</u> p, however, the expectation is not a trivial one. For the odd powers of i are either +i or –i. We have established that we can make this simple substitution, setting $x^p = X$, an integer, and $y^p = Y$, an integer. This gives us the expectation that:

$$X \pm Yi$$

is a hyper-volume measured by a real integer. As this is only possible if Y is zero—which disobeys the requirement that y be an integer greater than zero—such a combo is not possible. This holds for all odd primes, p, out to this side of infinity.

Waking Up

End of detour and back to multidimensional matrices.

A matrix in Hilbert space is the next step in sophistication. These are the tools with which to describe brain function; words are quite inadequate.

So, when you wake up in the morning, the virtual file stored on RNA is read back into active memory and resumes running in the VR generated by the brain. The "I Am" wakes up and heads for the bathroom. When running in the VR generated by my brain, this is the conscious "I."

In Volume Two: Science in the Realm of Spirit, we will look at the possibility of the same, self-assembled, labyrinthine "I Am" program running on a VR not generated by the physical human brain.

ME AND MY MAC

We can summarize this discussion with current, trans-millennial, computer science. For there is an almost perfect, one-for-one analogy—with a few 'minor' differences—between Me and my Mac. Bear me out, this is not just extreme Mac-love speaking. The differences to keep in mind while appreciating the analogy are:

Parallelism:

Mac: When not in 'sleep' mode and functioning efficiently, hundreds of regular programs run at a time in my Mac's active memory. A large percentage of these are on "idle" until they are called upon to do something.

Brain: When "I Am" is awake and functioning efficiently, a zillion programs are running in my brain's active memory. A percentage (estimated by some to be as high as 95%) of these are on "pause until they called upon by another program.

Operating system:

Mac: All these hundred-or-so programs are multi-threaded on 1 Motorola PowerPCG4 chip—an intricate assemblage of doped silicon—at a speed of 1,200,000,000 cycles/sec. This chip is running one legal copy of MacOSX.

Brain: Each of the zillion programs is single-threaded on a glial cell—an assemblage of CHNO—at a speed of scores of cycles/sec. Each of these zillion glial cells is running one copy of the brain's operating systems. The probably-incomplete list of these glial-run operating subsystems is: humanOS, hominidOS, mammalOS, reptileOS, fishOS, wormOS.

Program copies:

Mac: There is just one copy of each program running in active memory—such as Finder, Word, Photoshop, n-kernel, pbs, etc.

Brain: Each glial cell has a RNA linear subprogram that is just slightly different to that of the two glial cells on either side of it along a linear mental axis (a concept which we will shortly make mathematically rigorous). A glial cell can be a component of many such mental axes which thus cross in that cell. The six-degrees-of-separation rule probably holds, as billions of connected-glial lie along, and define, each active mental axis. Every one of these mental axes crosses the zero mental axis which is generated during development. This is equivalent to the unique zero point at the corner of a multidimensional cube in hyperspace. Each of the glial subprograms associated with an axis can be called upon by other RNA subprograms, from another axis, also in that glial. Thus, while there are a zillion programs running in my brain right now, there are only thousands of basically-different ones at work—each running in billions of slightly-different versions along each mental axis.

Connecting Buses:

Mac: The serial connections between central modules—the buses—are 64- or 128-bits wide, each bit being distinctive.

Brain: The serial connections between major modules—the white matter—are a zillion-bits wide, each bit being very slightly different to its neighbor's.

Long-term storage:

Mac: The contents of my Mac's active-memory—intricate longitudinal patterns of $+^{ve}$ and $-^{ve}$ electric charges in RAM—are constantly written and read, to and from, the short- and long-term sub-types of storage-memory. The short-term version is just another variety of RAM. The long-term version is quite different, it is a linear pattern of tiny N^{th} and S^{th} magnetic poles on a well-organized hard drive. Unused storage space on the hard drive is either blank or old stuff that can be written over. From the perspective of active-memory, however, both types of storage are identical except that short-term memory takes a few cycles to access while the long-term takes many, many cycles. There is a single hard drive which stores absolutely everything in long-term storage when I close the lid and the "Put to Sleep" program runs. This hard drive can be removed, plugged into another Mac and the active memory brought back to life, the lid can be opened, in a totally-new Mac.

Brain: The contents of my brain's active-memory—intricate transverse patterns of Na^+/K^+ membrane depolarization patterns in neural nets——are constantly written and read, to and from, the short- and long-term sub-types of storage-memory. The short-term version is just another variety of neural-net firing. The long-term version is quite different, it is a linear pattern of tiny A, U, G & C nucleotides on a memoryRNA subprogram in a glial cell. Unused storage space is always a blank—there is no such thing as writing over old memory, it is just added to. The moving finger writes in memoryRNA code, and it never overwrites, never even backspace-deletes. From the perspective of active-memory, however, both types of storage are identical except that short-term memory takes a few cycles to access while the long-term takes many, many cycles. Each glial cell stores such a memoryRNA subprogram of each of the mental axis it is a member of. This program is added to when that mental axis is in action. Each glial cell has stored absolutely everything in long-term storage when I am lying down, thinking vague thoughts of Oprah and tensors, and the "Put to Sleep" program runs. "I Am" when asleep is entirely stored as a billion-D real matrix of glial, with a set of memoryRNA programs at each location. All that really needs to be stored here are the values needed to specify a QPF, just q, p, I and s. As any three determine the fourth, only three values need to be kept in storage. Along the axis, these values would be the ones that alter a pixel at a time along the mental axis. This matrix of memoryRNA could theoretically be removed, dropped into another brain and the active memory brought back to life, the smell of good coffee is a good trigger, in a totally-new brain. Unless the brain is somehow a blank, I do not think this is nice idea, but it is theoretically possible. Subprograms of "I Am," such as knowing Hilbert-matrix-tensor algebra (see below for an introduction), would be a much better commercial prospect.

Programs running on a real operating system/chip:

Mac: The hundred-or-so programs running in active memory generate, in general terms, a single QPF that directs the functioning of keyboard, screen, or memory, etc. In the bottom-up perspective terms used in this book, this external activity is an example of stuff filling-in the QPF provided by programs. When I type a key on the Mac active and short-term memory spring into feverish activity. When I choose "Save," the long-term memory also gets involved.

Brain: When a glial cell runs one of its memoryRNA programs it generates a QPF that strongly influences the behavior of the local neural net it "serves". Each glial cell running a program generates such a QPF. This QPF, of course, obeys the equation $q=rI-rp$ As a mental axis is either active or inactive along its entire length, billions of similar programs, altering in pixel-by-pixel fashion the shape of QPF they generate along that axis. As thousands of higher programs, and billion sand billions of the lowest type, are all running—each on its own a mental axis—we end up with a zillion little QPF all interfering with each other. As the Mandelbrot set illustrated, in a very simple and 2-D, such mingling of complex numbers can result in interesting things. It is this final, massively-composite nervous system encompassing mindQPF that actually directs the firing of neural nets (even though the local glial have the most. When I type a key on the Mac, active-, short- and long-term memory all spring into feverish activity in my brain. The composite mindQPF around the finger changes and my finger moves to fill it in, hitting the key. From the bottom-up perspective, this is a flood of nerve impulse arriving in the muscles, and a flood of calcium ions making my actin molecules flip to a shorter shape. However, this is effect, the cause was the body-spanning composite mindQPF. There is no "save" button in my mind brain to call upon: Every thought, to a lesser extent, and every experience to an absolute extent, is reflected in the active and short-term memory as well as the endlessly extending memoryRNA program in each and every glial cell (many are blank, of course). This is the permanent hard drive storage where that implacable moving finger stores its, so far, unchangeable record of you and your life. The programs in this memoryRNA are what you and I really are. Some might thing this demeaning; I think it's brilliant! I'm delighted to know what I am. (It's better, to my mind's sense of dignity, than the 'you're a piece-of-meat, brain secretion' the classical science perspective suggests I am.)

Mobility:

Mac: The long-term storage form and transport is highly variable. This applies to all types of storage but we will use a program as an example. I can drag a copy of Word:Mac2004 from my hard drive to a virtual drive, from there drag a copy to a CD, from there grad a copy to replace the original on my Mac's hard drive (do not attempt at home). Same program, different external form. If I had downloaded this latest version of Word from the web over my cable wide-are LAN connection it would have been the same program but gone through the following forms as patterns of:

Pits and hills on the master DVD

Laser pulses

Electric charge on a Microsoft server's high-speed virtual hard drive

AC current in a laser driver

Light pulses in a coaxial cable

Repeat transformations ii. and iii many times

AC current in my cable modem, Ethernet cable and Airport module

Radio waves between Airport and Mac

AC current in my Mac's antenna

Electric charge in my RAM active- and short-memory

Patterns of Nth and S^{th} on my hard drive

Brain: Glial-matrix-memoryRNA is the only storage form we know of in which the human "I Am" comes in. Radio waves seem out of the question, though the reported 'golden cord' connection in out-of-the-body experiences is suspiciously like a computer bus or the corpus callosum.

This completes our list in which the computer Mac is not the same as the brain computer. For all that, they are minor differences compared to the low-resolution, general analogy we will to use in discussing the mind.

Keeping the limitations of the analogy in mind, we can now discuss how my brain is just like my Mac. Perhaps that is why we get along so well.

Consider this hierarchy of programs running on my Mac.

Word-for-Windows running on
  Windows NT running on
   VirtualPC running on
    Mac OSX running on the
     Unix shell calling
      Machine code controlling a
       Motarola chip.

WAKE UP!

I am typing away at the top, in Word-for-Windows, when I stop and start thinking about what I am writing. After one minute of inactivity, my Mac goes to sleep. It turns off the screen, writes a virtual image of the entire active- and short-term memory to the hard drive, halts the disk then turns everything off except a low-level little program that detects when a key is hit and calls the "wake up!" program. All this is to save battery power.

This file stores, all in one linear array of magnetic poles, an image of:

The state of Word-for-Windows + the state of WindowsNT + the state of VirtualPC + the state of Mac OSX + the state of the Unix shell + the state of the machine code

This is the Mac when it is asleep. We can call all this the magnetic image, m-image of the Mac in stored form.

Using this as an analogy, with all the limitations just listed, I can now describe myself at work on this book as:

The "I Am" User program running in a
  Virtual simulation of solid reality generated by the
    Virtualworld program running on the
      Brain OS running on the
        Module OS shell calling
          Neuron firing controlling a
            Body

While not as simple as the 'mind-and' concept, it is probably more accurate.

We can now use this to describe what happens when some low-level programs decides it is time to sleep and calls the "put to sleep" program from its RNA store along the glial on some mental axis.

There is no need to pause to "write to disk" as the glial have been doing this continuously all day. So active- and short-term memory can be just switched off, to clear them, and then filled with the nighttimes housekeeping chores program. Especially shoring up any new mental axes established during the day's experiences, like finding a long-elusive appreciation for rap music.

While I am asleep, the matrix of glial holds all of me in storage. This ultra-file stores, all in one 3-D array of memoryRNA, an image of:

The state of "I Am" + the state of Virtualworld + the state of the VR program + the state of the brain + the state of the modules + the state of the neurons

This is "I Am" when I am asleep. We can call all this the RNA image, the r-image of Me in stored form.

Waking up is so similar I will not discuss them separately. The image file is loaded back into active- and short-term memory, the neurons or RAM, and "I Am" and my Mac spring back to life to resume where we left of.

The human mind is actually programs running in a virtual reality. Philosophers talk about mental 'qualia,' such as 'red' or 'sweet." They are, of course, constructs of the VR. The program that generates the VR has a table that it compares the sensory input from the optic nerve, it stimulates the red mental axis appropriately. We only imagine we are experiencing a sweet red apple—it's really just an avalanche of neuron impulses. As we all inherited the same table from our distant ancestors, we all must 'see' the same red as the VR is the same for all of us. If this reminds you of The Matrix, you might have a point. (In mitigation, I believe that the purpose of it all is benign and beneficent, if given a chance, so Don't Panic.)

We stated that, if classical science is correct, this image of Me can only be contained in an r-image file. If most religions are correct, then we can assume that there is at least one other form that this Me image can be stored as.

Perhaps, like Airport and my Mac, they are as different as radio waves in a room and pits-on-a-DVD's silver surface are. We shall pick this up again in Volume Two.

BRAIN SCIENCE

Now for the concept of a mental axis and how it can be mathematically-described.

Each concept we have, each thing to which we give a name, a word. Each concept has a mental axis. Each concept has a word—or a structure of words that we use for it both inside our mind and when we speak externally of it.

Take the word 'quantum' as an example. You knew the word before you read this book so there was a mental axis assigned it, a linear chain of glial cells starting at the zero point zero.

If you are a well-read non-scientist, this was a meager chain, with few connections or intersections with other words and concepts.

If you are a specialist biologist, you started with a similar, if longer and better connected, mental axis for the word and concept "quantum."

If you are a quantum physicist, you will have started with a mega mental axis for the word with a mega number of intersections with other words and concepts each on its own mental axes.

If you have never come across the word quantum before, you started without a mental axis for it. If you have made it this far, however, a new mental axis will definitely have been extended and activated.

In all cases, hopefully, you will have made a lot of new connections and intersections between glial-stored mental axes.

We have already discussed all the math we need for basic brain science:

The glial cell array and its connections—the long-term storage mode—can be described This matrix is as a regular polygon in a multi-dimensional real space.

The active- and short term memory involves a QPF being generated at each entry, we have extension into complex space. This can be described with a multi-D Hilbert matrix with a q=rI-rp QPF at each entry. This matrix is as a regular polygon in a multi-dimensional Hilbert space and odd things can be expected to happen.

POPPER'S WORLD

We can conclude that Popper's three worlds are actually remarkably similar when looked at the quantum way we are suggesting:

World One, the physical universe. The material world is a composite of myriads of interfering QPF generated by protons & nuclei, filled-in by a myriad electrons.

World Three, the mental universe. The mind is generated by a composite of myriads of interfering QPF generated by RNA-programed glial cells, filled-in by a myriad neurons firing. There are many ideas: a few are allowed to 'escape' from the virtual reality

World Two, the cultural universe. There are many ideas in every human mind: a few of these ideas are allowed to 'escape' from the virtual reality and are expressed, for good or ill, in the external world. A culture is generated by a composite of myriads of these expressed QPF generated by the minds of the citizens, filled-in by a myriad of things such as families, audiences, law-courts, paintings, crime, corporations, concrete, wars, etc.

We conclude that, at least in this book, all three worlds are, basically, massively composite QPF being filled in by stuff. A theoretically Unified Science, indeed—all we need is a testable prediction...

Commuter's Fantasy

Did we forget anything? Oh yes, protein folding. Linear aminoacid chains finding their correct configuration; and calcium jumping.

Hopefully by now you have been so fully impressed with the power of quantum probability that that teleportation of aminoacid chains will pose no intellectual problem for it at all.

Consider a set up of two boxes separated by a good distance such that the probability of the test object being found in either box is 50% and being in-between is 0%. The experimenters regularly check the boxes. Half the time the object is in one box, half the time it is in the other. Yet, even if the boxes are miles apart, they are never found in transit between them.

Basic teleportation. (Actually, experimenters usually do the much simpler, if totally equivalent 'slit experiment,' but it's the same phenomenon.)

It takes a lot of ingenuity to demonstrate teleportation—the manipulation of quantum probability is still in its infancy—as things usually conspire to make the probability of moving incrementally exceedingly high and the probability of not doing so exceedingly small.

Starting with the simplest thing—the photon of light—experimentalists, over the years, have developed set-ups that have demonstrated the teleportation of electrons, atoms and even molecules—big ones.

I came across this with a google search:

The Vienna team, now jointly led by Zeilinger and Markus Arndt, has performed a new experiment on tetraphenylporphyrin molecules. These biological molecules are present in chlorophyll and hemoglobin. They have a diameter of about 2nm, which is over twice as big as a carbon-60 molecule.

These are big molecules the size of lipids, which is the polite term for fat and, as any dieter will attest, fat is decidedly flesh and "solid matter." Scientists are doing the equivalent of teleporting tiny bits of matter. The 'beam me up' days of commuter heaven will be here when these baby-step experiments are scaled-up somewhat. Don't burn your MetroCard yet, though, as the scale factor is about a trillion, trillion fold.

Now we can look at protein folding in terms of quantum probability forms instead of using the classical concept of lock-and-key.

Earlier we diagrammed the 'dissatisfied' aminoacids linked in the natal extended chain with their various needs and desires as the outies and innies of jigsaw pieces.

In the quantum view they are not 'solid' at all, the "bumps and holes" on them actually represent arrays of fully-filled, partially-filled and totally empty quantum probability forms. Now a 100-aminoacid chain is, admittedly, larger than a tetraphenylporphyrin molecule. But now the scale factor is a much more reasonable 50-fold

So, it is not unreasonable to suggest that a folding aminoacid chain takes the short cut and quantum jumps from the very-low probability extended state to the very-high probability, active form without trudging through everything in-between. Almost as rapidly, the chain could make a series of jumps to the final state.

Teleporting Proteins

Most of the interesting biochemical changes take energy input to make them happen, usually by involving ATP in the process. They are all endothermic energy-absorbing changes. Without the energy input from ATP, they would never happen even with all the correct enzymes present. Protein folding is almost unique in that it is an ectothermic process—it gives off heat and happens quite spontaneously.

So there just might be a way to test this. The extended chain has a higher free energy than the folded state. In the classical picture, this energy given out gradually and appears a slight raise in temperature. In the quantum view, this energy will be released all in one lump in the quantum jump and will appear as a photon of a specific frequency.

As the energy difference is not that great, the photons will be in the invisible infrared and microwave regions. I googled "visible light emission during protein renaturing" and got null results.

The outline of an experiment to test this is conceptually simple. Two cells—one the protein deficient control, the other a dilute solution of a simple and very pure enzyme—watched by sensors that span a wide frequency range.

We heat the solutions to above the denaturation temperature and then let them cool quickly. The sensors measure the spectrum of frequencies and make their report.

The classical prediction is that the thermal radiation of the test will show a simple shift compared to the control.

The quantum view suggests that there will be no shift in thermal spectrum but rather a spike, or series of them.

Heat denatures the chain, cooling renatures it. If quantum jumps are involved, then the opposite effect should also be observed: a laser tuned to the emitted wavelength should pop a folded chain back into the denatured state.

A few questions to conclude:

Can a tuned laser denature and renature enzymes?

Any entrepreneur have some spare venture capital to explore weightloss by teleport-o-liposuction?

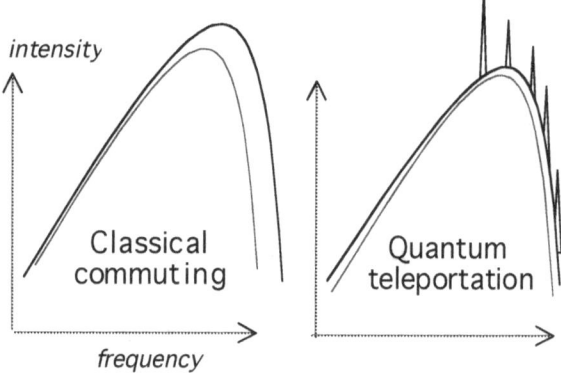

Summary

In Science, there is recognition of the universal impulse to choose a history that maximizes Quantum Satisfaction—a generalized inverse of the Principle of Least Action.

In a Quantum Science, natural law determines the Path of Greatest Satisfaction and thus the Quantum Probability of choosing to do something, or nothing.

Quantum Probability is very, very different to the chance-and-accident probability of classical science (and the chance of leaving Vegas with a cent).

Quantum Probability is all-powerful—from supporting old stars to the glint a diamond building to a baby's smile—and rules our universe with an inexorable, iron hand.

In a Quantum Science—other that the probability weighting just mentioned—all systems, from the simplest to the most sophisticated have true autonomy of choice. The only mathematical description of this is the True Random Number generator/operator (which, as such does not exist, does not help much). Moreover, things occasionally choose to do things with a vanishingly small Quantum Probability. This firmly- and painfully- established principle is summarized in the popular aphorism: Even God doesn't know which slit the electron will pick. Let alone me!

Therefore, while in classical science, the principle is that things must follow the Path of Least Action; in Quantum Science, things tend to follow the Path of Greatest Satisfaction.

The autonomy of all things is trumped by the well-accepted Law of Large Numbers. All this insists on is that, given enough repetitions, the actual history will faithfully flesh-out the Quantum Probability.

What things <u>are</u> involves long-term, interacting subsystems moving in a Quantum Probability Form (QPF). Our exemplar is the 1s (pronounced one-ess) orbital. We will also deconstruct and generalize the Schrödinger Equation to a form suitable for dinner conversation and T-shirts (see info for agent).

Reality is pixilated. Long-lasting QPF on each level of science are distinct and relatively small in number. Unlike classical science, which allows a continuum of forms, QPF are discrete and, at most as in the Human OS generated Virtual Reality, a denumerable infinity, not a continuum, is involved. Well-formed ideas are distinct.

What things <u>do</u> involves them exchanging body parts, with others of complementary tastes, via a Quantum Probability Field (also QPF).

One system (the generator) can provide a QPF for another, perhaps quite different, system to move in. Given sufficient time, it is inevitable that the filler-in system's history will reflect the QPF. Clay is our primitive example, proteins our most sophisticated.

Protein folding involves teleportation of the whole linear chain.

A triplet-code program, written on RNA, running once on the Basic OS 1.0 of life, generates a single metabolic QPF.

The Basic OS is described and its Evolution discussed.

Evolution occurs <u>first</u> on the internal, programming level. Novel programs are generated in a way reminiscent of antibody generation in lymphocyte.

There is a Virtual Reality QPF, generated by programs as above, in which programs are first tested before release. Errors are ruthless eliminated. Our exemplar is the lymphocyte and its testing at the hands of the thymus, a harsh master who kills those who fail his requirements.

I call this Internal Survival of the Well-formed, and it all occurs in the virtual, internal QPF VR.

Released into the real world to run, the program now faces the well-documented gauntlet of External Survival of the Fittest. This is the second, and quite subsidiary, step in evolution: the Darwinian reaper.

Variation is not random, it's stuff filling in distinct QPF and, in the long term, its destiny is utterly determined. (That should mollify the theologians for that seemingly-demeaning quip about God and the electron through two slits. They can claim God set up the probabilities, made some subatomic particles, then sat back to wait for the stuff to fill into the QPF and humans, at last, appear. Fear Not! For God will only appear, like this, in the text when I have a sidebar comment to inflame the to-the-death blood feud between classical science and traditional religion. As both have been superseded, however, they should take a time out for a breather, perhaps to watch the coronation of the Last Pope, Benedict-N, I will probably see.)

A bacterium has thousands of programs stored on its DNA. They are programmatically-called upon, as RNA transcripts, and the programs run in massive parallelism and repetition on millions of copies of the Basic OS. Each little program-run adds another QPF to the overall composite QPF.

The composite of billions of these QPF—interfering as waves in a wavefunction, a grand final probability amplitude with a tinge of Mandelbrot—is the QPF we recognize as a healthy, hearty bacterium when it is filled in by the atom-stuff. The QPF is relatively constant; the atoms are not, they are constantly replaced.

Metabolism can be usefully considered as flows of atom-stuff pouring in to fill in the QPF generated by the bacterial RNA programs. Then pouring right back out again. The timescale is longer, but the same is as true for us.

In brief, the massively-composite bacterial/organelle QPF is generated by many linear triplet-code RNA programs running massively in parallel on multiple Basic OS 1.0. Filled in by over time by a flow of atom-stuff, this filled-in QPF is what we call a healthy bacterium

Damage is simply healed by regenerating the QPF and waiting for the stuff to fill it in.

The elementary aspects of feelings and desires are discussed and established here on the bacterial level.

We now apply this basic concept to each of the levels in life's sophistication. In each case, we discuss the nature of each OS and its development in a womb-eden-mom-clay environment. On each level, healing and damage control is handled as above. Same for emotions. So:

Cell level: The massively-composite Cell QPF is generated by many linear spindle-code RNA programs running, massively in parallel, on multiple Cell OS 1.0. Filled in by over time by a flow of organelle-stuff, this filled-in QPF is what we call, and recognize, as a healthy eukaryote cell.

Organ level: The massively-composite Plant/Organ QPF is generated by many linear virish-code RNA programs running, massively in parallel, on multiple Organ OS 1.0. Filled in by over time by a flow of cell-stuff, this filled-in QPF is what we call, and recognize, as a healthy organ.

Body level: The massively-composite Body QPF is generated by many linear, higher-language RNA programs running, massively in parallel, on multiple Body OS 1.0. Filled in by over time by a flow of cell-stuff, this filled-in QPF is what we call, and recognize, as a healthy organ.

Basic Mind: The massively-composite mind of a worm is generated by many linear glial-code RNA programs running, massively in parallel, on multiple Neural-1 OS 1.0. This is filled in by over time by a flow of patterns of neuron firing.

This is expressed via the nerves in the well-established way. This filled-in QPF is what we call, and recognize, as a healthy and happy worm.

Fish Mind: The massively-composite mind of a fish is generated by many RNA programs running, massively in parallel, on multiple Fishy OS 1.0. This is filled in by over time by a flow of patterns of neural net firing. This filled-in QPF is what we call, and recognize, as a healthy and happy fish.

Amphibian Mind: The massively-composite mind of a fish is generated by many RNA programs running, massively in parallel, on multiple Fishy OS 1.0. This is filled in by over time by a flow of patterns of neural net firing. This filled-in QPF is what we call, and recognize, as a healthy and happy turtle.

Reptile Mind: The massively-composite mind of a reptile is generated by many RNA programs running, massively in parallel, on multiple Dino OS 1.0. This is filled in over time and is what we would have called, and recognized, as a healthy and happy dinosaur.

We conclude that, yes, proteins do teleport in an RNA world.

The task of science is to deconstruct all these RNA-bourn programs on RNA, or dePrograming as I like to call it.

Etc.

Time Frame. The rule is that internal and external evolution of programs can only proceed until they become subprograms called by a higher language. An OS rarely changes, and then by just a tweak.

Basic OS 1.0 emerged 4.2 billion years ago in a black-smoker per fused China clay bed; the first eden-womb.

Cell OS 1.0 emerged some 3 billion years ago in a womb-eden stromatolite.

Organ OS 1.0 emerged in the Ocean womb-eden some billion years ago.

The animal body-basic brain or Fish OS emergence kicked off the Cambrian Explosion some half billion years ago. The VR testing place was perfected as the Basic Mind we also have as a part of our Mind.

Then the Amphibian OS 1.0 emergence and its subsequent VR development and perfection as the basic Amphibian Mind we also have in our composite Mind.

Same for the Reptile OS 1.0.

And the Mammal OS 1.0.

The Primate, Ape and Hominid VR sequentially emerged with increasing sophistication until, less than a 100,000 years ago, the VR was perfected as the Human OS, the Mind in which you, a Master Program are running and trying to make sense of all this. What you think of as thinking is actually a program running in a VR, which is, to my mind. Somewhat less demeaning than being told I'm just a bunch of neurons sparking.)

The emergence of any new sister species, and in particular, the first emergence of the Human Mind VR in Adam and Eve involves:

The tetraplex of meiosis and the little-understood recombination complex. It is in the tetraplex of the male that a previously unoccupied QPF gets filled-in with a quantum jump (or 'the Word of God' takes form, as the religionists would have it).

Four generations:

A pre-grandfather in whose testis the empty Human VR generating program is first filled in. His offspring are the:

Hominid pre-parents, who carry the Human Program as a massive, inactivated Barre Body in their germ cells. Parents of:

The first True Humans, born from a hominid womb with regular navels. The language instinct emerges about age four and culture begins.

The birth of humans in the regular way. A new species, but much more significantly, a new VR, has successfully evolved internally, and the become established externally.

Diploid sperm, triploid zygotes, four-ploid germ cells and other such oddities.

The nature of sleep, dreaming and Dreaming is discussed. This ends with a teaser for Volume Two, which deals with the Planck Mirror that separates the physical and spiritual realms.

APPENDIX

1. Complex numbers

Not only is the sequence of cause-and-effect more sophisticated in quantum physics; so is the math.

In grade school we are taught the law of signs which might as well end up with: Minus times minus is a plus, for reasons we will not discuss. It really is tough to make sense of this rule using regular numbers.

Fortunately, mathematicians, starting in the Renaissance, came to accept that the math of the regular numbers (the one we learn in school) is incomplete. Such real numbers, as they are called, cannot deal, for instance, with the square root of negative numbers, and these square roots pop up all over in pure and applied mathematics.

The completion of the number realm used in math was the expansion of the domain of numbers into the imaginary and the complex.

In essence, while regular numbers allow for a measure of the size of things, complex numbers measure a size and direction at the same time. You might usefully recall a concept mentioned earlier at this point—that the cause-of-probability has size and direction.

The real numbers, the familiar ones, actually do have a direction to them. But there are only two of them—a direction of 0°, which is what the positive numbers have, and a direction of 180° which is the direction of the negative numbers going in the opposite direction.

Complex numbers are basically the same as these real numbers; the only difference is that they can have any angle of rotation from 0° to 360°.

When you multiply numbers with a direction you add their angles together. With this, the explanation of 'minus times minus is a plus' is as simple as two half-rotations bring us back to where we started:

$$180° + 180° = 360° = 0°$$

It is instructive to remember how difficult the concept of negative numbers seemed until quite recently. Take five oranges away from four oranges. How many oranges are left? Weird question. Eventually mathematicians realized that allowing for negative numbers introduced no contradictions, and in fact empower their calculation skills.

In general, numbers-with-direction have a size or magnitude, p, and an angle or amplitude, a.

number-with-direction = p@a

(Note that zero can be considered either to have no direction at all or all directions at the same time—it ends up the same thing).

From +1 and –1 you can easily construct all the real numbers such as 2, 3, 3.5, 4 etc., and -2, -3, -3.5, -4, etc.

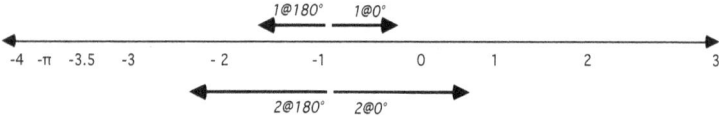

RULES

The basic rules for manipulating numbers-with-direction are very simple:

to add, put the numbers head-to-tail

to multiply, add the angles and multiply the sizes

Subtraction and division are just the inverse of these.

With this definition of multiplication, and being aware that a full rotation of 360° about the origin brings us back to 0°, we see the emergence of the rules for multiplying positive and negative numbers.

Positive numbers have 0° direction. When multiplying them, we add the two zeros together to get zero, so multiplying two positives—adding two zero angles—results in a size with zero direction, a positive number.

+1 x +1 = (1@0°) x (1@0°) = (1 x 1) @(0° + 0°)

= 1@0° = +1

Negative numbers have a 180° direction. Adding 180° to 180° gets us to 360°, all the way around to zero. So multiplying two negative numbers results in a size with 0° direction, again a positive number.

−1 x −1 = (1@180°) x (1@180°) = (1 x 1) @(180° + 180°)

= 1@0° = +1

Multiplying a positive and a negative—adding 0° to 180°—results in a size with 180° direction, a negative number.

+1 x −1 = (1@0°) x (1@180°) = (1 x 1) @ (0° + 180°)

= 1@180° = −1

This is how modern math deals with plus and minus numbers. It is perhaps not obvious, at this point, just what advantages this number-with-direction viewpoint has over more simple ways of dealing with positive and negative numbers. Yet this simple-to-grasp perspective will help immensely in comprehending the complex numbers with all sorts of directions.

The positive and negative "line" of numbers are all lumped together as the real numbers—the ones that lie on the axis through the zero point.

These real numbers are quite sufficient, and eminently useful, for measuring the external aspects of the world which is why we learn about them in elementary school. To be sure, the term "real" is in the Platonic sense as numbers, as entities, are actually quite abstract. To a mathematician at least, the number "two" has an abstract existence that is independent of "two things". Recognition of abstract entities is more difficult than concrete ones, and, for all the "obvious" utility of the real numbers in describing many of the quantitative aspects of our world, seeing that "two sticks" and "two daughters" had something in common took time and was an historical advance. The number "zero," the last integer to be recognized, wasn't fully acknowledged until the twelfth century.[1]

The fruit of all this effort, however, was most constructive as the real numbers bear their title well; they are eminently suited to describing the quantitative way in which many real things behave.

IMAGINARY NUMBERS

While it had appeared briefly in earlier mathematics, it was only after the Renaissance that mathematicians finally confronted the fact that the mathematics of the real numbers was incomplete. Simply put, the real numbers were incapable of dealing with the square-roots of negative numbers (let alone their cube-roots, etc.).

Now finding the square-root of a number is considered an elementary operation in math. They are as common, and as important, as are addition and multiplication. For example, the final step in solving the fundamental Pythagoras equation involves finding the square-root.

As often as not, however, in solving their equations, mathematicians ended up with negative numbers under the square-root sign. If they were lucky, these "unsolvable, meaningless, imaginary" numbers would cancel out, and a solution could be obtained.

---

1 Georges Ifrah, From One to Zero, Viking, 1985.

As often as not, however, such cancellation did not occur, and the equation was deemed unsolvable or as having an "imaginary" solutions (a put-down that was later adopted.)

One thing is clear, the square-root of a minus number cannot be any of the real numbers, for both positive and negative ones give positive numbers when squared.

The solution is simple in hindsight. Once we allow for rotation of numbers by units of 180° we can start thinking about numbers with a direction that is not 0° or 180° but something else.

We are looking for a number-with-direction that, when multiplied by itself, gives a negative number, a number-with-direction with an angle of 180°.

When we multiply two numbers-with-direction we add the two angles and multiply the two sizes (always a real, positive number). We want to solve this equation for the two unknowns:

p@a x p@a = p²@2a = 1@180°

There are just two solutions:

1@90° x 1@90° = 1@180°

1@–90° x 1@–90° = 1@180°

These two numbers are so useful that they have their own symbols, i and –i. The imaginary numbers as they are called in that they hover directly above and below zero on the real line.

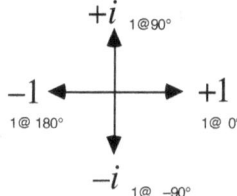

Just as +1 and –1 are the basis for all the infinity of the real numbers, so +i and –i are the basis for an infinity of imaginary numbers such as 2i, 1/3i, 3.5i, 4i etc. and -2i, -1/3i, -3.5i, -4i etc.

It was these imaginary numbers, with directions +90° and –90°, that allowed for solutions to the mathematicians dilemma.

LITTLE ARROWS

From here, it is no big leap for us to consider numbers that are not limited to having a directions just multiples of a right angle from 0°. What about numbers with any size and any direction? These are called the complex numbers and are central to the way the new physics describes the world.

We, of course, have the benefit of hindsight—the leap to these "complex numbers" took many a genius years of struggle to get the concept clear. A complex number can have any size, any direction. We will use the symbol p for the size of a complex number and a for its direction—and diagram it as an arrow with size p and angle a. It is these "little arrows" that feature so prominently in the math of quantum physics.

Complex numbers were discovered by mathematicians long before their remarkable usefulness in physics was understood. They are now as useful as the real numbers—inasmuch asmost of the fundamental equations of 20th century science use complex numbers. Both scientists and technologists would be lost without them—try understanding quantum mechanics or AC circuits without them—it's totally impossible. Like trying to do sophisticated arithmetic without a zero.

Just for completeness, we shall briefly look at the various forms that complex numbers take, each with its particular usefulness.

We have already mentioned the polar forms, describing the arrow in terms of size and direction. The rectangular form is the two components of the arrow, its projections on the real and imaginary axis—a combination of a real and an imaginary number.[1]

$z = p e^{i\pi a} = p@a$ polar form

$\quad = x + yi$ rectangular form

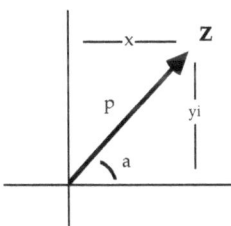

A complex number is, in the terminology discussed earlier, a 1D-extension in a 1D Hilbert space, the simplest of all.

The most common form in quantum physics is the exponential form. This is even more so for electronic AC theory where, somewhat confusingly, they use little-j to signify the square-root of minus-one. This tradition started because little-i was already firmly-established as signifying the negative intensity of the electric current. So j is used:

$z = p e^{j\pi a}$

where 'e' is the 'natural' exponential base, a real, transcendental number that starts 2.17…. and 'a' is measured in radians, not degrees. Technically, as an angle greater than 360° restarts the numbers at 0, this 'a' is actually, (a mod 2π) radians.

This is a natural extension of the concept of the exponential function: when you raise a number to a real power you alter its size; when you raise a number to an imaginary power you alter its rotation.

The two forms are related by Pythagoras:

$p^2 = x^2 + y^2$

Also trigonometry:

$\tan a = y/x$

Mathematicians, scientists and technicians always use radians. This 'natural' measure of angle is based on there being 2π radians in a full circle of rotation. The conversion table is:

Positive real: $a = 0 = 360° = 2\pi$

Negative Real: $a = 180° = \pi\pi$

$90° = \pi/2$

So all the following expressions are equivalent and interchangeable:

$-1 = 1@180° = -1 + 0i = e^{ji\pi}$

$\sqrt{-i} = 1@90° = 0 + 1i = e^{ji\pi/2}$

We shall only explore the very fringe of complex number math—just sufficient to appreciate how they so-perfectly describe the cause-of-probability in the new physics. The first thing is that complex numbers come in 'families' of four. They all have the same magnitude, it is the angles that relate them. This "family" of four related little arrows often appear together in quantum descriptions.

the number $p@a$

the negative $p@a+180°$

---

1 E. Mayor, The Story of a Number.

We mentioned earlier that, unlike +1 and –1, the twins +i and –i are virtually indistinguishable. This is why they often appear together in equations. The conjugate of a complex number simply replaces i with –i. Equivalently, put a minus sign in front of the angle. So the conjugates of the twins above are:

the conjugate p@ –a

the negative conjugate p@ 180°– a

While this sounds complicated, the diagram shows how simple their relations are. Think of the horizontal real axis as a mirror. Then the conjugate is just the "reflection" of the number in this mirror. Similarly, the negative conjugate is its reflection in the imaginary, vertical mirror. And the negative is the reflection of both conjugates in both mirrors.

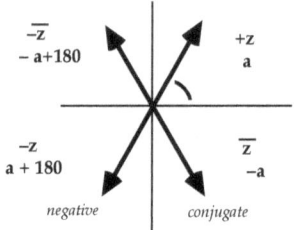

While the properties of the complex numbers are fabulous and enthralling to the mathematician, the only complex math we will use as examples will involve this simple family of complex numbers combining with each other in different ways.

For example, demonstrating yet again what Nobel laureate Eugene Wigner called "the unreasonable effectiveness of mathematics in the natural sciences," a combination of complex number and conjugate describes exactly the connection between the internal and external aspects of the new physics while the combination of a complex number with its negative describes how "nothing" can shield a target from a "something."

Key Properties

Complex numbers are able to completely describe all the properties of the probability amplitudes and quantum cause-of-probabilities in general. Complex numbers are "unreasonably effective" in describing the internal extension of matter. The properties of complex numbers fundamental to quantum physics are multiplication, addition and conjugation. Other behaviors, such as subtraction and division of complex numbers, are defined by mathematicians, but are not needed in describing the behavior of the quantum cause-of-probability.

Addition: Adding complex numbers is most simply accomplished in the rectangular formation—just add the real and imaginary components separately.

$$x+yi \; + \; u+vi \; = \; (x+u) \; + \; (y+v)i$$

This is equivalent to putting the arrows head to tail in a diagram—the result is the arrow that joins the start and finish. The size of the final arrow depends on the angles—adding size 2 to size 2 can give a variety of sizes, including size 4.

Adding a complex number to its negative is adding plus and minus equal quantities so the result is zero:

$$(x+yi) \; + \; (-x - yi) \; = \; (x-x) + (yi-yi) \; = \; 0$$

This exact canceling of adding a number and its negative will prove important when we describe combinations that have exactly zero resultant size.

Multiplication: Multiplication of complex numbers is simplest in the arrow, or polar, formulation—an arrow with a size and a direction. We have already encountered this rule: the magnitudes (sizes) multiply each other while the angles (amplitudes) sum together.

$$p@a \times q@b = pq @ a+b$$

It is impossible to have a probability greater that 100% even in the quantum world. For this reason, in quantum physics the size of the arrows is never greater than unity. So multiplication either leaves the size alone—if the multiplicand has size exactly one—or it is less-than-one and so shrinks the final size. Thus the occasional reference to multiplication in QED as shrink-and-turn of little arrows.

Multiplication of complex numbers is used in modern physics to describe how sequences of cause-of-probabilities combine with each other.

Collapse to real: The last key property is that of multiplying a complex number with its twin, its conjugate.

$$p@a \times p@-a = p \times p @ a-a = p^2$$

Almost by definition, the angles always cancel out so this operation always gives a real, positive number with an direction of exactly 0°. This combination of complex numbers is called the absolute square. We will refer to this as this the "squaring" of a complex number into a regular number with just size but no direction—a scalar, as the mathematicians would have it.

We will usually symbolize the absolute square of a complex number with P (think probability) or by any of the other, all equivalent, representations.

Two things are of note here.

Cancellation: If a QPF and its negative are added together, the size of the result is exactly zero. If p is zero, P is zero. The probability is zero; it is forbidden. This is what underlies the bullet-proof nothing we encountered earlier.

If $p = 0$ then $P = 0$

Internal Amplification: If we double a QPF, we double its size while the angle stays the same. Doubling the size of the probability amplitude quadruples the probability:

$$z = p@a \; 5p@a \; 10\,p@a$$

$$z^2 = P\; 25\,P\; 100\,P$$

We will refer to this as internal amplification of probability. It this that underlies the high probability of the 'contented' state of paired electrons that drives chemical interaction. Clearly, if we are dealing with a gazillion amplitudes adding, the amplification of probability that results is going to be a gazillion-squared. This is exactly what happens in a laser where the probability of all the photons jumping into exactly the same state is so overwhelmingly-large that they all jump there at the same time and a laser beam of coherent light zaps out. This is why it is called Light Amplification by Stimulated Emission of Radiation—a phenomenon Einstein predicted long before it was actually observed.

2. Slit Experiment

We will now take a look at the experiment that played a pivotal role in the quantum revolution, the slit experiment, the equivalent of our weird execution. This experiment actually incorporates almost every aspect of quantum weirdness.

Much of the early history of experiments that lead to quantum mechanics involved trying to figure out if the basic stuff of matter was made of particles or waves. In classical science there is a clear distinction between a particle and a wave. A particle stays together as it moves through space while a wave spreads out.

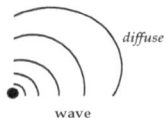

In classical physics there was a simple "slit" experiment that could tell if something was a particle or a wave. The essentials are simple: fire the thing at a barrier with holes in it and watch what happens.

We can predict what happens when we fire particles at a barrier by considering a cannon firing balls at a wall in which there is a slit. We expect to see balls imbedded in the walls and a pile of balls on the other side that made it through the hole. The first key point is that each ball will arrive at a certain location on the far side.

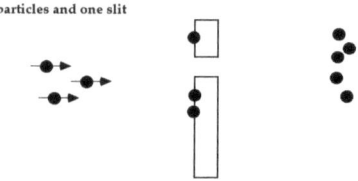

We repeat the experiment after opening another slit in the wall.

We now expect to find that more particles have traveled through the two holes combined.

We do not expect that opening another slit will prevent balls making it through the first slit to the other side.

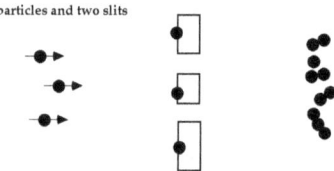

We expect something quite different to happen when we aim a wave at the slits—perhaps an ocean lapping against a wall with inlets to waters beyond. With one slit open the waves pass through and we expect to see the wave energy deposited in a diffuse zone on the further shore. This is quite different from the particles which arrive at specific locations.

With two slits open we expect something called "interference" to happen. On the far side of the barrier there are now two waves, and the crests and troughs of one wave will overlap those of the other.

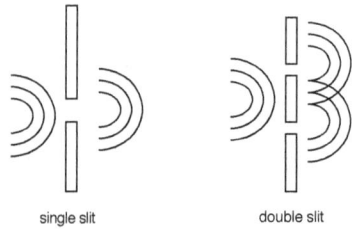

When the crests and troughs are "in phase" they will constructively interfere and combine their effects into extra big crests and troughs. When they are "out of phase," they interfere destructively and can even cancel each other out—there is no wave there at all.

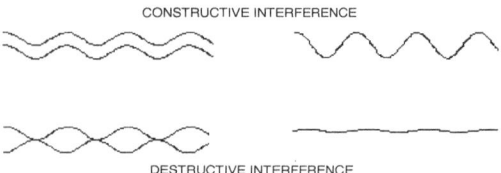

The end result of this interference is that we expect the wave detectors on the far side of the barrier to register places where there is constructive interference—and much energy is deposited—and places where destructive interference occurs and little energy arrives there. When waves are involved, we will not be surprised if opening two slits register zero at a detector that fires when either slit is open.

Opening both slits creates a destructive interference at the detector and it registers no energy arriving in the wave.

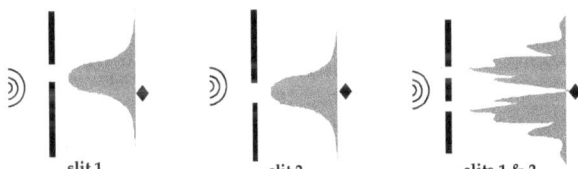

With waves, we can expect "nothing" to block them from reaching certain places. This is quite different what we expect to happen when particles are being projected at single and double slits.

PARTICLE OR WAVE

The slit experiments were thought to be a simple way to distinguish between particles and waves because the predictions are so clear-cut.

	Worldline	Interference
wave	diffuse	constructive and destructive
particle	distinct	none

The distinction is very clear and scientists used such slit experiments to answer such questions as: Is light a wave or a particle? Is an electron a particle or a wave? Is an atom a wave or a particle?

It is with such slit experiments that physicists attempted to answer the question: Is light a particle phenomenon or a wave phenomenon?

Since the 1800s, it was known that when light passes through narrow slits it exhibits wave-like properties—it spreads out and exhibits interference patterns.

In experiments with light there were detectors that fired when either was open but not when they were both open.

The results eemed clear-cut—light was a wave.

For many years, such experiments were taken as convincing evidence that light was a wave phenomenon. Early physics had this wave occurring in an aether that pervaded all space; later thinking replaced this with undulation in an almost-equally-enigmatic electromagnetic field.

Newton had suggested that light was composed of particulate "corpuscles", but "regular" particles would certainly not be expected to behave in such a fashion. Opening both slits should make it easier for more particles to get from one side to the other. If light were classical particles, the expectation would be that more of them would get through when both slits were open than would get through when just one or the other slit was open.

Such behavior, however, can easily be explained by interference, and the consensus for many years was that light is a wave.

Compelling evidence gradually accumulated, however, that light could not be just a continuous wave phenomenon because it behaved, in many situations, as if it was composed of discrete particles, now called photons. For instance, when photons travel singly through the apparatus they behave as particles and fire a single detector—they do not arrive diffusely as we expect a wave to do.

Another example is that a high-energy photon can bounce off an electron, just like two pool balls colliding (the Compton Effect).

The upshot of this and many other experiments is that the photon has to be considered just as much a particle as is an electron.

Light, it seemed, was both a particle and a wave and, for a period, this dichotomy engendered various explanations of light such as a wave-particle wavicle—a dual-natured thing that was simultaneously a particle and a wave—or, even more contrived, that the instruments used to examine light determined whether you saw a particle or a wave.[1]

This fuzzy thinking soon became unnecessary. Explaining the slit-experiment pattern with just waves became untenable when sophisticated single-photon detectors were developed (just a little better than the three-photon sensitivity of the human eye). Exactly the same pattern could be produced over a long period of time when just one photon at a time passed through the apparatus. In the simple two-slit experiment, a single photon passing through the apparatus can fire the detector when either slit is open but never when both are open.

Here light is behaving as a particle at the beginning and end of the experiment—the emitter and the detector deal in single photons—and as an interfering wave while going through the apparatus—a single photon interferes with itself.

The basic systems of matter seem to have both particle and wave aspects. We will shortly see that the wave aspect is the internal wavefunction—hence the name—while the particle aspect is the external structure and interactions of the system.

The slit experiment has one more surprise to offer the theoretician. One obvious way to figure out if a photon is passing through the slit experiment as a particle or a wave is to put little detectors in the slits. Detectors that tell you if a photon passed through their slit.

The good news is that the experiment has been done and, yes, photons do pass through one slit or the other—i.e. as particles—and not through both as would waves. The bad news is that somehow, no matter how subtle the detectors, the characteristic pattern of the interference is no longer there. The photons behave as regular particles and arrive just as little cannon balls would be expected to.

Somehow, putting in the slit detectors makes this a quite different experiment, one that involves only particles behaving in a classical way.

PARTICLE AND WAVE

Coming from the other end of the "wavicle spectrum," similar interference patterns can be created in experiments with the decidedly-particulate electron—in either the all-together or the one-at-a-time situations. The electron can also behave as a particle or as a wave.

In the spring of 1991, four different laboratories independently demonstrated the interference of atoms which are indisputably bits of matter. "The first to report was Professor Jürgen Mlynek.... The sketch of [his] apparatus might have come from Young's own papers: the experiment itself was a repetition of the original 1803 version, with the crucial difference that the slits were irradiated not by sunlight but by a stream of material particles.... The most mysterious feature of the experiment... is the fact that each atom traversed the apparatus alone, uninfluenced by the jostle of other particles."[2]

---

1 Isaac Asimov, The History of Science, Walker & Co. NY (1985), pp. 378-9.

2 Hans Christian von Baeyer, Taming the Atom: Emergence of the Visible Microworld, Random House, NY (1992), pp. 166–7.

This is, as noted, equivalent to the teleportation of stuff that is decidedly matter—scaling this up a zillion times, that is.

Clay

Clay molecules are excellent catalysts, almost as versatile as platinum at providing paths of least resistance for other systems. The energy—the external enabler—is provided by nature in the form of the UV-driven iron cycle (which only shuts down with the advent of significant oxygen production by photosynthesis) and the smokers pumping out high-energy sulfides etc. (which they are still doing, and powering a bizarre ecology).

The clay provides the wavefunction down which these high-energy systems interact with each other. This is a vertical provision in that systems on different levels can be manipulated by clay: atoms, phosphate, carbohydrates, aminoacids, etc.

The clay provides a path of least resistance for reactants to turn into products. In classical terms, the catalyst is said to "stabilize" the intermediate by lowering its free energy. Now while this might seem like a rather complex way of looking at catalysis compared to the classical view of a lock-and-key where the reactants "fit" onto a surface and get stabilized, we shall see that this way has far more explanatory power when we get to more sophisticated levels.

Even today, all metabolic transformations involve catalysis by surfaces. Nowadays all these surfaces are provided by the endlessly versatile proteins, but biogenesis must have involved much simpler surfaces as proteins, themselves, can only realistically be created by a metabolism more capable that just nature-in-the-raw.

The systems that we know emerged with catalytic ability in the proto-metabolic era are many and various. Examples are the iron sulfides—can energize molecules from inorganic sources—and the clay molecules that are versatile in providing QPF for reactants to transform into product.

The black smokers under the oceans are prodigious providers of activated iron sulfides; and great beds of clay are a historical relic of this anchient period of time.

One reviewer of the current understanding of the origin of life concluded, "The most reasonable interpretation is that life did not start with RNA [DNA is not even under consideration as it is even more sophisticated than RNA]. The RNA world came into existence after many of the problems associated with prebiotic synthesis and template-directed replication of RNA had been solved. This implies that there was a simpler genetic system, or systems, that preceded RNA and that the evolutionary advances made by the ancestral system were somehow carried over to the RNA world."[1]

One of the most compelling suggestions as to what systems were involved in pre-life manipulation of molecules is the thesis, developed by Dr. Cairns-Smith, that primitive life first emerged in clay and clay structures.[2]

Dr. Cairns-Smith makes a good case in demonstrating that it is much more plausible to assume that simple systems, what he calls 'low tech,' emerged first and provided a foundation upon which more sophisticated 'high tech' systems could develop. He proposes that a 'low–tech' manipulation of molecules developed before the 'high–tech,' remarkably-complex manipulation of molecules by the gene/protein triplet code system.

This would be a forgotten sub-basement in the skyscraper of genetics.

While there is evidence that triplet code-based foundations of all life on Earth was established by 3.5 BYP. There is also evidence that life was making an impact on the earth even before this at almost 4 BYP[3]. The problem with both protein and nucleic acid polymers as ingredients of the earliest life has been called the 'Uroboros Puzzle' (the mythical serpent with its

---

1 Gerald F. Joyce, "RNA evolution and the origins of life," Nature 338, 1989, pp. 217-224.

2 Cairns-Smith, A. G., (1982). Genetic Takeover And The Mineral Origin Of Life, Cambridge University Press.

3 Manfred Schidlowski, "A 3,800-million-year isotopic record of life from carbon in sedimentary rocks," Nature 333, 1988, p. 313.

tail in its mouth): To make proteins, nucleic acids are required: to make nucleic acids, proteins are required—a chicken-and-egg type of can't-have-one-without-the-other conundrum. "This is the essence of the Uroboros problem."[1]

Dr. Cairns-Smith makes a compelling case for clay being the low-tech, pre-life provider of wavefunctions—he doesn't use this terminology, of course. He even goes so far as to imagine quite complex structures of multiplication and natural selection. On the other hand, clay might just as well have indulged in organic chemistry just for the hell of it, and living systems gradually took over the basic manipulation of molecules. The actual history was probably a mix of these two extremes.

This wavefunction provided by, say, clay—the classical catalytic "surface"—is a subset of the wavefunction of the clay system wavefunction. Earlier, we went to great pains to show that empty wavefunctions are as objectively-real as filled wavefunctions. Much of the catalytic activity of platinum, for instance, can be ascribed to the many empty orbitals just beneath its surface that provide a temporary home—a path of least resistance—for coupling electrons that are otherwise unable to get over a bump in the road.

---

1 Ronald F. Fox "Energy and the Evolution of Life" 1988, p. 35.

www.ingramcontent.com/pod-product-compliance
Lightning Source LLC
Chambersburg PA
CBHW080915170526
45158CB00008B/2115